AF664044

ALGEBRA

ALGEBRA

Sudarsan Nanda
Professor of Eminence and Research Chair
KIIT Deemed to be University
Bhubaneswar 751024, Odisha, India
Retired Professor of Mathematics, IIT Kharagpur
Former Vice Chancellor, North Orissa University

ALLIED PUBLISHERS PVT. LTD.

New Delhi • Mumbai • Kolkata • Chennai • Bangalore • Hyderabad

ALLIED PUBLISHERS PRIVATE LIMITED

D-5, Sector-2, **Noida**–201 301
Ph. Nos.: 0120-4320295/2542557/4352866 • E-mail: delhi.books@alliedpublishers.com

17 Chittaranjan Avenue, **Kolkata**–700072
Ph.: 033-22129618 • E-mail: cal.books@alliedpublishers.com

15 J.N. Heredia Marg, Ballard Estate, **Mumbai**–400001
Ph.: 022-42126969 • E-mail: mumbai.books@alliedpublishers.com

No. 25/10, Commander-in-Chief Road, Ethiraj Lane (Next to Post Office)
Egmore, **Chennai**–600008
Ph.: 044-28223938 • E-mail: chennai.books@alliedpublishers.com

P.B. No. 9932, No. 15, 3rd Floor (Next to Vijaya Bank), 5th Cross, Gandhinagar, Karnataka,
Bangalore–560009, Ph.: 080-41530285 / 22386239
• E-mail: bngl.journals@alliedpublishers.com / apsabng@airtelmail.in

Sri Jayalakshmi Nilayam, No. 3-4-510, 3rd Floor (Above More Super Market)
Barkatpura, **Hyderabad**–500027
Ph.: 040-27551811, 040-27551812 • E-mail: hyd.books@alliedpublishers.com

Website: www.alliedpublishers.com

ISBN: 978-93-90951-23-9

Published by Sunil Sachdev and printed by Ravi Sachdev at Allied Publishers Pvt. Ltd.,
D-5, Sector-2, Noida–201 301

Dedicated

to

my teacher
Professor Gokulananda Das,
my maternal uncle Sri Banshidhar Mishra,
Professor Achyuta Samanta,
my family Kuni, Mamuni, Rony & Rosy
and
to my students.

Preface

This book is suitable as a text on algebra and linear algebra for BA/BSc Mathematics students of all Indian Universities and second and third year Undergraduate students of the Universities of North America and Europe. This is prepared according to the model syllabus of UGC which has been accepted by several Universities of India including all the Universities of Odisha state.

There are several excellent books on algebra. However, there are not many titles by Indian publishers or low priced reprints of foreign publishers, which are exactly according to the syllabi of Indian Universities. Some of them are very elementary, long and elaborate, while some are advanced and too compact and hence become difficult to grasp by our undergraduate students. Some books also contain many additional topics which are not included in the syllabus. Therefore, the author felt the need for writing a book which is lucid and simple, neither too compact nor too elementary, that would contain the basics of a first course and also a number of examples to illustrate the abstract concepts.

Chapter 1 is preliminaries, chapters 2–5 are the topics for Abstract Algebra and Chapters 6 and 7 are the topics of Linear Algebra as per the model syllabus of UGC. Chapter 9 is an additional topic on Fuzzy Set Theory for the interested students.

The book is the first of its kind for various reasons. First, it contains the theory of fuzzy sets. Fuzzy set theory is of fairly recent origin, first introduced by Professor L.A. Zadeh in 1965, which has found applications in various branches of knowledge. No book on algebra published so far contains fuzzy sets. Second, the book has been written in such a manner that students pursuing courses in Number Theory and theory of Equations can appreciate how abstract concepts have application in these fields. Chapter 8 a small chapter on Boolean Algebra has been included since some Universities may include this in their course.

The prerequisites are a course in Higher-Secondary Mathematics. However, the book is self-contained with a preliminary chapter on set theory. This course is prerequisite for advanced courses on abstract algebra. Real analysis, topology, functional analysis and linear algebra.

Previously these topics were taught to M. Sc students of Indian universities and in the recent past, even before the introduction of three-year degree courses, they figured in the curriculum of BA/BSc. The contents of this book have been taught by the author to the students of REC (NIT) Rourkala, IIT Kharagpur, Utkal University Bhubaneswar, and KIIT Deemed to be University, Bhubaneswar.

During preparation, I was greatly benefitted and have made use of the works of several authors including the book entitled 'Topics in Algebra' by I. N. Herstein. A complete list appears in the section of Bibliography. I take this opportunity to acknowledge my indebtedness to them.

The manuscript is prepared when I am at Kalinga Institute of Industrial Technology (KIIT) Deemed to be University, Bhubaneswar. I express profound thanks to Professor Achyuta Samanta, Founder KIIT and KISS and the authorities of KIIT for cooperation, encouragement and help. This book is exactly as per UGC model syllabus and the syllabi of Universities of Odisha and the chapters have been arranged to maintain the sequence in the syllabus. I thank Dr. Srikumar Acharya for formatting, editing and proof reading of the manuscript and Allied Publisher Private Limited for publishing the book.

I have taken enough care to make the book free from errors; despite all my efforts, some of the mistakes might have survived. I will be grateful to readers if they are kind enough to point them out. Finally I must express that the purpose will be best served if it benefits students for whom it is meant.

Sudarsan Nanda

Contents

CHAPTER 1

Preliminaries

The axiomatic method of Mathematics which started with Euclid's Elements revived in the end of nineteenth century and made enormous progress since the beginning of twentieth century. Cantor (1845–1918) was the first to systematically study the theory of sets and hence, regarded as the creator or the father of set theory. He published his papers on Set Theory mostly during the period 1874–1897. Cantor tried to define a set. His definition was "A collection into a whole of definite, well distinguished objects of our intuition or thought". In his paper (3) Cantor believed.

"There exists a set A containing those and only those elements which satisfy any prepositional function ϕ".

The earlier attempt to axiomatize Cantor's Set Theory was made by G. Ferge. He included the so called axiom of abstraction which asserts the existence, for any given property *P*, of a set whose members are precisely those objects having the property.

Mathematicians did not work long with ideas of Cantor and Ferge before a crisis developed. One of the first and the simplest of such contradictions was made by Bertrand Russell in 1903 which is called Russell's Paradox.

This paradox led to various axiomatizations of Set Theory. The two types of collections: arbitrary collections in the sense of Cantor which we call *classes* and certain special type of collections that we call *sets* were distinguished. Russell and Whitehead [15] resolved this difficulty by discovering theory of types. They established a hierarchy of types of collections: A collection x can be a member of another collection y only if y is one level higher in the hierarchy than x. X is class if it is a collection of objects satisfying some prepositional function. Such a collection may not be a set.

Godel-Bernays system (see Bernays [1]) discussed the theory of classes which is assumed to have three primitives (undefined concept): sets, classes and membership. Zermelo-Frankel Theory (see Zermelo [20]) discussed set theory which is assumed to have two primitives: sets and membership. Thus, we do not define a set and an object or member of set.

1.1 SETS AND SET INCLUSION

We usually denote sets by capital letters such as *A, B, C,* .... and elements of a set by small letters such as *a, b, c,* .... To indicate that an element *a* is a member of a set *A,* we write,

$$a \in A$$

which is read '*a* is an element of *A*' or '*a* belongs to *A*'. To indicate that the element *a* is not a member of the set *A*, we write,

$$a \notin \mathrm{A}$$

which is read '*a* is not an element of *A*' or '*a* does not belong to *A*'.

There are two standard notations available for designating a particular set. One way of describing a set is to list the elements of the set between braces. Thus, the set of all positive integers that are less than 5 can be written as:

$$\{1, 2, 3, 4\}$$

the set of all positive integers can be written as,

$$\{1, 2, 3, \ldots\}$$

the set of all odd integers can be written as,

$$\{\pm 1, \pm 3, \pm 5, \ \ldots.\}$$

and the set of all fourth roots of unity can be written as,

$$\{1, i, -1, -i\}$$

With this notation an arbitrary set with six members may be written as,

$$\{a, b, c, d, e, f\}$$

At times it is inconvenient or impossible to list all the elements of a set. In such cases, a set can often be described by specifying a property that the elements of the set have in common. For example, consider the set,

$$A = \{1, 2, 3, 4, 5, 6\}$$

which has been described here by listing the elements.

A is precisely the set of all positive integers that are less than 7 and this can be written as:

$$A = \{x \mid x \text{ is a positive integer less than } 7\}.$$

which is read '*A* is the set of all *x* such that *x* is a positive integer less than 7.' As another example consider the following expression,

$$\{x \mid x \text{ is real and rational}\}$$

which is read 'the set of all *x*, such that *x* is real and rational, denotes the set of all real numbers which can be written as the quotient of two integers. The four sets of numbers described at the beginning of this paragraph can be written either way:

$$\{1,2,3,4\} = \{x \mid x \text{ is positive integer less than } 5\}$$
$$\{1,2,3,\ldots.\} = \{x \mid x \text{ is a positive integer}\}$$
$$\{1, \mathrm{i}, -1, -\mathrm{i}\} = \{x \mid x \text{ is a complex number and } \mathrm{x}^4 = 1\},$$
$$\{\pm 1, \pm 3, \ldots\} = \{x \mid x \text{ is an odd integer}\}.$$

Two sets A and B are equal if and only if they have the same elements. This is known as the *axiom of extension.* In this case we write,

$$A = B$$

If A and B are not equal, then we write,

$$A \neq B$$

For example, if

$$A = \{1,2,3\}$$

And

$$B = \{x \mid x \text{ is a positive integer and } x^2 < 16\}$$

Then

$$A = B.$$

Sometimes one sees sets that have no members. For example, the sets

$$\{x \mid x \text{ is rational and } x^2 = 2\},$$
$$\{x \mid x \text{ is real and } x^2 = -1\}$$

have no members. A set with no members is called an *emptv* set or a null set and such a set is denoted by the symbol ϕ. A set is said to be *finite* if it is empty or consists of n elements for some positive integer n; otherwise it is said to be infinite.

More precisely the existence of the empty set follows from the *axiom of specification* which says that 'To every set A and to every condition $S(x)$ there corresponds a set B whose elements are exactly those elements x of A for which $S(x)$ holds'. By applying this axiom to any set A with the condition '$x \neq x$' we get the empty set,

$$\{x \in A : x \neq x).$$

If every element of a set A is also an element of set B, then we say that A is a subset of B or is contained in B and write,

$$A \subseteq B$$

We sometimes express this by saying that B is a superset of A (or contains A). $A \subseteq B$ allows for the possibility that A and B might be equal. If A is a subset of B and A is not equal to B, then we say that A is a proper subset of B or is properly contained in B and this is written,

$$A \subset B.$$

We can also express this by saying that B is a proper superset of A or properly contains A. The existence of subsets follows from the axiom of specification. It is customary to agree that empty set ϕ is a subset of every set.

Example 1: If

$$A = \{a, b, c\}$$

then all possible subsets of A are,

$$\Phi, \{a\}, \{b\}, \{c\}, \{a, b\}, \{b, c\}, \{a, c\}, \{a, b, c\}.$$

Example 2: If

$$A = (x \mid x \text{ is a positive integer}\},$$

$$B = \{x \mid x \text{ is an integer}\},$$

then

$$A \subset B$$

It is often convenient to have a symbol for logical implication. If p and q are statements, then $p \Rightarrow q$ means that p implies q, or that if p is true, then q is true. Similarly, $p \Leftrightarrow q$ means that p is equivalent to q, or that p is true if and only if q is true.

If A is any set, then

$$A \subseteq A$$

Let A, B and C be three sets. Then

$$A \subseteq B \text{ and } B \subseteq A \Rightarrow A = B,$$

$$A \subseteq B \text{ and } B \subseteq C \Rightarrow A \subseteq C.$$

Note that,

$$A = B \Leftrightarrow A \subseteq B \text{ and } B \subseteq A.$$

If a set A is not a subset of B we write,

$$A \not\subset B.$$

1.2 ALGEBRA OF SETS

The intersection of two sets A and B is the set of all elements those belong to both A and B, it is denoted by the symbol,

$$A \cap B.$$

If $A \cap B \neq \phi$ we say that A intersects B. If $A \cap B = \phi$, then we say that A does not intersect B or that A and B are disjoint. That $A \cap B$ is a set follows from the *axiom of specification* if we write,

$$A \cap B = [x \in A : x \in B\} \text{ or } A \cap B = \{x \in B : x \in A\}$$

The intersection of any collection of sets is the set of all elements those belong to every one of the sets in the collection.

Example 3: Let

$$A = \{1, 2, 4, 7, 11\}$$
$$B = \{2, 7, 9, 11, 13, 21\}$$
$$C = \{0, 2, 5, 7, 17, 23\}$$
$$D = \{2, 7, 13\}$$

Then,

$$A \cap B \cap C \cap D = \{2, 7\}.$$

The operation of forming intersections is commutative and associative:

$$A \cap B = B \cap A$$

and

$$A \cap (B \cap C) = (A \cap B) \cap C.$$

It also has the property that,

$$A \cap A = A, A \cap \phi = \phi.$$

The union of two sets A and B is the set of all elements those belong to either A or B or both, it is denoted by the symbol,

$$A \cup B.$$

The above definition can also be expressed as,

$$A \cup B = \{x | x \in A \text{ or } x \in B\}.$$

The operation of forming unions is also commutative and associative,

$$A \cup B = B \cup A$$

and

$$A \cup (B \cup C) = (A \cup B) \cup C.$$

It also has the properties that,

$$A \cup A = A, A \cup \phi = A.$$

The union of any collection of sets is the set of all elements that belong to one or more of the sets in the collection. That it is a set follows from the

Axiom of union: For any collection of sets there is a set that contains all the elements that belong to at least one set of the given collection.

Example 4: $A = \{1, 2, 4, 7, 9, 17\}$

$B = \{2, 7, 9, 11\}$

$C = \{0, 17\}$

$D = \{1, 7, 13\}$

Then,

$$A \cup B \cup C \cup D = \{0, 1, 2, 4, 7, 9, 11, 13, 17\}.$$

Note that the idea of using shading und circles to illustrate the relationship between sets is attributed to the British logician John Venn and in his honour, they are called Venn diagrams.

We have now defined two of the fundamental operations on sets and we shall now sec how they are related to each another. The facts here are given by the distributive laws:

$$A \cap (B \cup C) = (A \cap B) \cup (A \cap C)$$

$$A \cup (B \cap C) = (A \cup B) \cap (A \cup C)$$

We now give a logical proof of the distributive laws. We prove the first distributive law and leave the second as an exercise.

Let $x \in A \cap (B \cup C)$

$\Rightarrow$ $x \in A$ and $x \in B \cup C$

$\Rightarrow$ $x \in A$ and $x \in B$ or $x \in A$ and $x \in C$

$\Rightarrow$ $x \in A \cap B$ or $x \in A \cap C$

$\Rightarrow$ $x \in (A \cap B) \cup (A \cap C)$

Hence,

$$A \cap (B \cup C) \subseteq (A \cap B) \cup (A \cap C)$$

Let $x \in (A \cap B) \cup (A \cap C)$

$\Rightarrow$ $x \in A \cap B$ or $x \in A \cap C$

$\Rightarrow$ $x \in A$ and $x \in B$ *or* $x \in A$ and $x \in C$

$\Rightarrow$ $x \in A$ and $[x \in B \text{ or } x \in C]$

$\Rightarrow$ $x \in A \cap (B \cup C)$

Hence,

$$(A \cap B) \cup (A \cap C) \subseteq A \cap (B \cup C).$$

Thus

$$A \cap (B \cup C) = (A \cap B) \cup (A \cap C).$$

This proves the first distributive law.

The next of our major operations on sets is the formation of complements.

Given any set A, it may be necessary to consider those elements that are not in A. Usually, however, there are elements not in A that are extraneous to the problem under consideration. For example, let A be the set of all positive integers. Clearly, if x is a negative integer, then $x \notin A$. However, if x is a cabbage then x is also not in A since cabbages are not positive integers. If we are concerned with a problem about integers then 'an element x not in A' is really meant 'an integer x that is not in A'. Cabbages are extraneous to the problem.

One way of eliminating extraneous elements is to specify in advance that all elements in the problem under consideration come from some fixed set U, called the universal set. For example, in a problem concerned with real numbers, we might agree to take the universal set U to be the set of real numbers and by 'an element x that is not in A' we must mean 'a real number x that is not in A'. In many problems, the universal set is clear from the nature of the problem and is not explicitly stated.

If U is the universal set and A is a subset of U, then the set of all elements in U that are not members of A is called the complement of A; it is denoted by the symbol A'.

The operation of forming complements has the following obvious properties:

$$(A')' = A, U' = \phi, \phi' = U.$$

$$A \cup A' = U, A \cap A' = \phi.$$

Example 5: Let

$$U = \{0, 2, 9, 17, 23, 37\}$$

And, $V = \{0, 17, 23\}$

Then, $V' = \{2, 9, 37\}$

The following two important results, called *De Morgan's laws,* give relationship of complementation with union and intersection.

If A and B are any two sets, then

$$(A \cup B)' = A' \cap B'$$

$$(A \cap D)' = A' \cup B'.$$

Example 6: To illustrate De Morgan's law

Let $U = \{a, b, c, d, e, f\}$

$A = \{a, b, d\}, B = \{b, d, f\}$

Then $A \cup B = \{a, b, d, f\}$
$(A \cup B)' = \{c, e\}$
$A' = \{c, e, f\}, B' = \{a, c, e\}$
$A' \cap B' = \{c, e\}.$

We now prove the first law of De Morgan's law

Let $x \in (A' \cap B')$
$\Leftrightarrow$ $x \in A'$ and $x \in B'$
$\Leftrightarrow$ $x \notin A$ and $x \notin B$
$\Leftrightarrow$ $x \notin A \cup B$
$\Leftrightarrow$ $x \in (A \cup B)'$

Hence $A' \cap B' = (A \cup B)'$

This completes the proof.

We now introduce the difference and symmetric difference of two sets. If A and B are two sets we define,

$$A - B = \{x \in A : x \notin B\}$$

That this is a set follows from the axiom of specification. It is easy to see that,

$$A - B = A \cap B'.$$

We define the symmetric difference of two sets *A* and *B* as,

$$A \Delta B = (A - B) \cup (B - A)$$

This operation is commutative and associative, i.e.

$$A \Delta B = B \Delta A$$

$$A \Delta (B \Delta C) = (A \Delta B) \Delta C.$$

Also, $A \Delta \phi = A, A \Delta A = \phi$

It can be checked easily that,

$$A \Delta B = (A \cup B) - (A \cap B).$$

Given a set *U* and an element *u* if we use the digits 0 and 1 to denote $u \notin U$ and $u \in U$ respectively, then we can represent membership in tabular forms in the following manner.

Membership Table for $\boldsymbol{A'}$	
A	A'
0	1

Membership Table for **A ∪ B**		
A	B	$A \cup B$
0	0	0
0	1	1
1	0	1
1	1	1

Membership Table for **A ∩ B**		
A	B	$A \cap B$
0	0	0
0	1	0
1	0	0
1	1	1

We have been considering the subsets of a given universal set U; do those subsets themselves constitute a set? The following axiom guarantees that the answer is yes.

Axiom of Powers: For each set there exists a collection of sets that contains among its elements all the subsets of the given set.

In other words if U is a set, then there exists a set $P(U)$ such that if $X \subset U$, then $X \in P(U)$.

The set P described above may be larger than wanted; it may contain elements other than the subsets of U. This is easy to rectify; just apply the axiom of specification to form the set $\{X \in P(U): X \subset U\}$. Since, for every X, a necessary and sufficient condition that X belongs to this set is that X be a subset of U, it follows that if we change notation and call this set $P(U)$ again, then

$$P(U) = \{X: X \subseteq U\}$$

This set P(U) is called the power set of U; the axiom of extension guarantees its uniqueness.

Because the set $P(U)$ is very big in comparison with U, it is not easy to give examples. If $U = \phi$ then,

$$P(A) = \{\phi\}, P(\{a\}) = \{\phi, \{a\}\} \text{ and}$$

$$P(\{a, b\}) = [\phi, \{a\}, \{b\}, \{a, b\}].$$

The power set of triplet has eight elements. The reader can probably guess (and is hereby challenged to prove) the generalization that includes all these statements: the power set of a finite set with n elements has 2^n elements.

1.3 CARTESIAN PRODUCT AND RELATIONS

In many situations we are interested in paired data. An ordered pair (a, b) is a listing of two objects a and b in definite order. The element a is called the first entry and the element b is called the second entry of the ordered pair. Two ordered pairs are said to be equal if they list the same objects in the same order. Thus, for (a, b) and (c, d) to be equal, we must have $a = c$ and $b = d$.

Thus two ordered pairs (a, b) and (b, a), though involve the same elements, convey different information because of the order unless $a = b$.

We are now in a position to define the Cartesian product of sets. If A and B are two sets, the Cartesian product of A and B is the set of all ordered pairs (a, b) where $a \in A$ and $b \in B$, it is denoted by $A \times B$.

Thus,

$$A \times B = \{(a, b) | a \in A, b \in B\}$$

Example 7

If $\quad A = \{a, b, c\}$

and $\quad B = \{1, 2\}$

then $A \times B = \{(a, 1), (a, 2), (b, 1), (b, 2), (c, 1), (c, 2)\}$.

It is possible to define the Cartesian product of more than two sets. For example, if $A_1, A_2 \ldots A_n$ are n sets, where n is a positive integer, then the Cartesian product,

$$A_1 \times A_2 \times \ldots \times A_n$$

is defined by,

$$A_1 \times A_2 \times \ldots \times A_n = \{(a_1, \ldots, a_n) | a_1 \in A_1, \ldots a_n \in A_n\}.$$

A relation is a set of ordered pairs. Thus, a relation is any set which is a subset of the Cartesian product of two sets. Let X and Y be any two sets (which may be equal, but need not be) and let R be a relation from X into Y. Then $R \subset X \times Y$. We define the domain of R to be the set dom $R = \{x \in X: (x, y) \in R \text{ for some } y \in Y\}$ and the range of R to be the set range $R = \{y \in Y: (x, y) \in R \text{ for some } x \in X\}$. The symbol R^{-1} denotes the inverse of R: $R^{-1} = \{(y, x): (x, y) \in R\}$.

There are two types of relations which often arise in mathematics: equivalence relations and order relations.

1.3.1 Equivalence Relation

Definition: Let X be a nonempty set. An equivalence relation '$\sim$' on X is a relation such that for $x, y, z \in X$ we have,

1. $x \sim x$ (reflexive);
2. $x \sim y \Rightarrow y \sim x$ (symmetric);
3. $x \sim y$ and $y \sim z \Rightarrow x \sim z$ (transitive)

An equivalence relation on X has an intimate connection with decompositions of X into nonempty subsets which fill it out and have no elements in common with one another.

A partition of X is a disjoint class $\{X\}$ of nonempty subsets of X whose union is the full set X itself.

In other words, a partition of X is the result of splitting it, or subdividing it, into nonempty subsets in such a way that each element of X belongs to one and only one of the given subsets.

We now show that each partition of X has associated with it a natural equivalence relation in X and conversely, a given equivalence relation in X determines a natural partition of X.

Let X be a nonempty set. Assume that a partition of X is given. We now associate with this partition a relation '$\sim$' in X. We say that $x \sim y$ if x and y belong to the same partition set. It is easy to see that $\sim$ defines an equivalence relation in X.

Conversely let $\sim$ be an equivalence relation in X. Let x be an arbitrary element of X. The set defined by $[x] = \{y : y \sim x\}$, which is the set of all elements of X which are equivalent to x, is called the equivalence set of x. We show that the class of all distinct equivalence sets forms a partition of X. By reflexivity $x \in [x]$ for each $x \in X$, so each equivalence set is nonempty and their union is X. It remains to show that any two equivalence sets $[x_1]$ and $[x_2]$ are either disjoint or identical. Suppose that $[x_1]$ and $[x_2]$ are not disjoint. To be precise, let $z \in [x_1] \cap [x_2]$. Now $z \sim x_1$ and $z \sim x_2$ and by symmetry $x_1 \sim z$. Let $y \in [x_1]$, so that $y \sim x_1$ now since $y \sim x_1$ and $x_1 \sim z$ transitivity shows that $y \sim z$. Since, $y \sim z$ and $z \sim x_2$ again transitivity shows that $y \sim x_2$ so $y \in [x_2]$. Since y is arbitrary element of $[x_1]$, this shows that $[x_1] \subseteq [x_2]$.

The same reason shows that $[x_2] \subseteq [x_1]$.

Thus, $[x_1] = [x_2]$ and this completes the proof.

We now consider several important and simple examples of equivalence relations.

Example 8: Let I be the set of all integers. For $a, b \in I$ we define $a \sim b \Leftrightarrow a = b$.

Then it is evident that $\sim$ defines an equivalence relation in the set I and each equivalence class consists of precisely one integer.

Example 9: Let D denote the set of all fractions, i.e. let D be the set of all symbols of the form a/b, where $a, b \in I$ and b is not zero. If a/b and c/d are two fractions, then we define a relation $\sim$ in D as follows:

$$\frac{a}{b} \sim \frac{c}{d} \Leftrightarrow ad = bc.$$

Then for each $\frac{a}{b}, \frac{c}{d}, \frac{e}{f} \in D$.

We have

1. $\frac{a}{b} \sim \frac{a}{b}$;
2. $\frac{a}{b} \sim \frac{c}{d} \Rightarrow \frac{c}{d} \sim \frac{a}{b}$;
3. $\frac{a}{b} \sim \frac{c}{d}$ and $\frac{c}{d} \sim \frac{e}{f} \Rightarrow \frac{a}{b} \sim \frac{e}{f}$.

Thus $\sim$ defines an equivalence relation in D and each equivalence class of fractions is what we call a rational number.

1.4 FUNCTION

Let X and Y be any two sets (which may be equal but need not be). A function from X into Y is a rule f which assigns to each element $x \in X$ unique element $y \in Y$. The y which corresponds in this way to a given x is usually written $f(x)$, and is called the image of x under the rule f, or the value of f at the element x. A function f can also be viewed as a relation, that is, a subset of $X \times Y$ in which $(x, y) \in f$ and $(x, z) \in f$ implies that $y = z$.

A function f from X into Y is often called a mapping, or a map, or a transformation from X into Y. The set X is called the domain of f, written as dom f, and the set of all values of f for all x in X is called its range and is written as rng f. A function whose range consists of just one element is called a constant function.

If f is a function from X into Y it is denoted by the symbol $f : X \to Y$.

If it is clear from the context what the sets X and Y are, or if there is no need of specifying them explicitly, it is a common practice to identify the function $f: X \to Y$ by the rule f without mentioning the sets X and Y. If two definite sets X and Y are under discussion, then a mapping of X into Y is sometimes designated by

$$x \to y$$

where, y is the image of x. For example, the function $f: R \to R$ where R denotes the set of all real numbers defined by,

$$f(x) = x^2$$

can be designated by,

$$x \to x^2$$

A function f is called an extension of a function g (and g is called a restriction of f) if the domain of f contains the domain of g and $f(x) = g(x)$ for each x in the domain of g.

If

$$f: X \to Y, \text{ let}$$
$$G_f = \{(x, y) \in X \times Y : y = f(x)\}.$$

This set G_f is called the graph of f. The graph G_f is related to the function f in the same sense that either one of them completely determines the order.

Consider a mapping $f: X \to Y$. Then clearly range of f is a subset of Y, that is rng $f \subseteq Y$. If $f: X \to Y$ is such that range of f equals Y, then we call f surjective, or a mapping of X onto Y. If any two distinct elements in X have distinct images under f, then we call f injective, or a 1–1 mapping from X into Y. If f is both injective and surjective, then we call f bijective, or a one-one correspondence from X onto Y. If $f: X \to Y$ is bijective, then we can define its inverse mapping $f^{-1}: Y \to X$ as follows: for each $y \in Y$ there exists a unique element $x \in X$ such that $f(x) = y$ since f is 1–1 and onto; we then define x to be $f^{-1}(y)$.

Let X be a nonempty set. The identity mapping on X is the mapping of X onto itself which leaves each element of X fixed.

Suppose that $f: X \to Y$ and $g: Y \to Z$ be any two mappings. The product or the composition of these two mappings, denoted by $gf: X \to Z$, is defined by,

$$(gf)(x) = g(f(x)).$$

We observe that the two mappings involved here are not arbitrary, for the set Y which contains the range of the first, equals the domain of the second. More generally, the product of two mappings is meaningful whenever the range of the first is contained in the domain of the second.

Example 10: $f: R \to R$ be denoted by,

$$y = f(x) = x^2 + 1$$

and

$g: R \to [-1, 1]$ be defined by,

$$g(x) = \sin x.$$

Then

$gf: R \to [-1,1]$ is defined by,

$$(gf)(x) = g\big(f(x)\big) = g(x^2 + 1)$$
$$= \sin(x^2 + 1).$$

Consider an arbitrary mapping $f : X \to Y$. If $A \subseteq X$, then its image $f(A)$ is the subset of Y defined by,

$$f(A) = \{f(x) | x \in A\}.$$

Similarly if $B \subseteq Y$, then its inverse image $f^{-1}(B)$ is the subset of X defined by,

$$f^{-1}(B) = \{x | f(x) \in B\}.$$

The first mapping has the following properties:

$$f(\phi) = \phi; f(X) \subseteq Y;$$
$$A_1 \subseteq A_2 \Rightarrow f(A_1) \subseteq f(A_2),$$
$$f\left(\bigcup_{i \in I} A_i\right) = \bigcup_{i \in I} f(A_i)$$
$$f\left(\bigcap_{i \in I} A_i\right) \subseteq \bigcap_{i \in I} f(A_i).$$

The second one is much better behaved:

$$f^{-1}(\phi) = \phi, f^{-1}(Y) = X,$$
$$B_1 \subseteq B_2 \Rightarrow f^{-1}(B_1) \subseteq f^{-1}(B_2)$$

$$f^{-1}\left(\bigcup_{i\in I} B_i\right) = \bigcup_{i\in I} f^{-1}(B_i)$$

$$f^{-1}\left(\bigcap_{i\in I} B_i\right) = \bigcap_{i\in I} f^{-1}(B_i).$$

1.5 COUNTABLE SETS

Let A and B be any two sets. If there is a function from A onto B which is also one-one, then f is called a one-one correspondence between A and B. If there is a one-one correspondence between two sets A and B, then we say that A and B are numerically equivalent or that A and B have the same cardinal number and we write it as $A \sim B$.

The relation '$\sim$' of numerical equivalence between two sets is in fact an equivalence relation. It can be easily verified that

1. $A \sim A$ for any set A, i.e., '$\sim$' is reflexive,
2. If $A \sim B$, then $B \sim A$, i.e. '$\sim$' is symmetric,
3. If $A \sim B$ and $B \sim C$, then $A \sim C$, i.e., '$\sim$' is transitive.

A set is called finite if it is either empty or is numerically equivalent to the set $\{1, 2, 3, \ldots\ldots, n\}$ for some natural number n. A set A is called infinite if it is not finite. The simplest and most common example of an infinite set is the set N of all natural numbers:

$$N = \{1, 2, 3, \ldots\ldots\ldots\}$$

A set A is said to be countable if it is finite or is numerically equivalent to N. An example of a countable set is the set A of all even positive integers. In this case the function $f: A \to N$ defined by,

$$f(n) = \frac{n}{2}$$

is a $1-1$ correspondence. Another example is the set A of all integers. In this case a $1-1$ correspondence $f : N \to A$ is given by,

$$f(n) = \begin{cases} \dfrac{n}{2} & (n \text{ even}) \\ \dfrac{n-1}{2} & (n \text{ odd}) \end{cases}$$

EXERCISES

1. Let $f: X \to Y$ be an arbitrary mapping. Define a relation in X as follows: $x_1 \sim x_2$ means that $f(x_1) = f(x_2)$. Show that this is an equivalence relation and describe the equivalence sets.
2. In the set R of all real numbers, let $x \sim y$ mean that $x - y$ is an integer. Show that this is an equivalence relation and describe the equivalence sets.
3. Let I be the set of all integers, and let m be a fixed positive integer. Two integers a and b are said to be congruent modulo m symbolized by $a \equiv b \pmod{m}$ if a – b is exactly divisible by m, i.e, if $a - b$ is an integral multiple of m. Show that this is an equivalence relation, describe the equivalence sets.
4. Decide which ones of the four properties reflexivity, antisymmetry, symmetry, and transitivity are true for each of the following relations in the set of positive integers: $m \leq n, m < n, m$ divides n, $m = n$.
5. Let X be a non-empty set. A relation $\sim$ in X is called circular if $x \sim y$ and $y \sim z \Rightarrow z \sim x$, and triangular if $x \sim y$ and $x \sim z \Rightarrow y \sim z$. Prove that a relation in X is an equivalence relation $\Leftrightarrow$ it is reflexive and circular $\Leftrightarrow$ it is reflexive and triangular.
6. Give an example of a relation which is (a) reflexive but not symmetric or transitive; (b) symmetric but not reflexive or transitive, (c) transitive but not reflexive or symmetric; (d) reflexive and transitive but not symmetric; (f) symmetric and transitive but not reflexive.
7. Let X be a non-empty set, and let f and g be one-to-one mappings of X onto itself. Show that fg is also a one-to-one mapping of X onto itself and that $(fg)^{-1} = g^{-1}f^{-1}$.
8. Let f, g and h be any three mappings of a non-empty set X into itself, and show that multiplication of mappings is associative in the sense that $f(gh) = (fg)h$.
9. Let X and Y be non-empty sets and f a mapping of X into Y. Show the following:
   (a) If f is one to one $\Leftrightarrow$ there exists a mapping g of Y into X such that,
   $gf = idx$
   (b) f is onto $\Leftrightarrow$ there exists a mapping h of Y into X such that $fh = idy$.
10. Let X be a non-empty set and f a mapping of X into itself. Show that f is one-to-one onto $\Leftrightarrow$ there exists a mapping g of X into itself such that $fg = gf = idx$.

CHAPTER 2

Integers and Congruence

2.1 INTEGERS

Oldest and most basic mathematical system is the set N of all positive integers. Kronecker once said "God made the positive integers; all else is due to man". However, it was shown by the Italian Mathematician G. Peano (1858–1932), that the positive integers can even more simply be defined by a set of axioms, known as Peano axioms. To describe these axioms we first need to define the so-called Peano's successor function. This function, denoted by s, assigns to every positive integer the next larger one. Thus $s: N \to N$ is defined by,

$$s(n) = n + 1.$$

In terms of the function s, a number of basic properties can be formulated which characterize the set N. These properties can be summarized as follows:

(a) $1 \in N$

(b) $n \in N \Rightarrow s(n) \in N$

(c) $\exists$ no $n \in N$ such that $s(n) = 1$

(d) $s(n) = s(m) \Rightarrow n = m$

(e) if $A \subset N$ such that $1 \in A$ and $s(n) \in A$ whenever $n \in A$, then $A = N$.

Axiom (e) leads to one of the most important principles of Mathematics, known as the principle of mathematical induction.

We write N for the set of all positive integers and Z for the set of all integers.

Given $a, b \in Z\,(b \neq 0)$ if there is $c \in Z$ such that $a = bc$, then we say that b divides a and we write $b \mid a$. It is clear that,

(i) If $c|b$ and $b|a$, then $c|a$

(ii) If $c|b$ and $c|a$, then for any $m, n \in Z$ we have $c \mid ma + nb$.

Positive integers can be divided into two classes: Prime numbers and Composite numbers.

Definition

$p \in N$ is said to be a prime number if:

(i) $p > 1$

(ii) p is not divisible by any positive integer except 1 and p.

On the other hand, any positive integer greater than 1 which is not a prime is called a composite number.

Theorem 2.1.1

Every positive integer greater than 1 is the product of (certain) prime numbers.

Proof: Suppose that the given number is c. If it is a prime number then there is nothing to prove. If it is a composite number then it is divisible by some integers greater than 1 and less than c. Let p_1 be the smallest of such integers. Then p_1 is a prime number. For if p_1 is divisible by some positive integer $l > 1 (l \neq p_1)$ then l will also divide c. But l being less than p_1 this contradicts the choice of p_1. Thus c is divisible by a prime number p_1. Let $c = n_1 p_1$. Here $1 < n_1 < c$. Now n_1 is either a prime or is divisible by a prime number p_2. i.e., $n_1 = n_2 p_2$ where $1 < n_2 < n_1$. Similarly, n_2 is either prime or is divisible by a prime number p_3. Thus $n_2 = n_3 p_3$ where $1 < n_3 < n_2$. Continuing in this way we get a decreasing sequence $\{n_1, n_2, n_3, \ldots\}$ of positive integers each greater than 1 and less than c. Let n_k be the smallest of these. Now n_k is a prime number $(= p_k)$ for otherwise if n_{k+1} divides n_k then n_k will not be the smallest of $\{n_1, n_2, n_3, \ldots\}$. Hence $n = p_1\ p_2\ p_3 \ldots p_k$ and this completes the proof.

Note that the prime numbers $p_1, p_2, p_3, \ldots$ need not be different from each other. If some of the p_m's are identical, then c will be represented as $c = p_1^{a1} p_2^{a2} \ldots p_m^{am} (a_1 > 0,\ a_2 > 0, \ldots, p_1 < p_2 < \ldots)$

We now prove.

Theorem 2.1.2 (The fundamental theorem of Arithmetic)

The above representation of c is unique.

Before proving this we first prove.

Theorem 2.1.3 (Euclid's First Theorem)

If some prime number p divides ab, then p divides either a or$b$.

Proof: Suppose that p does not divide a.

We now show that p divides b.

Firstly, consider the set $(pX + aY$: X, Y are integers). Clearly, $pX + aY$ is an integer whenever X and Y are integers. Let q be the least positive integer which can be expressed in the above manner. So $q = px_1 + ay_1$ for some

$x_1, y_1 \in Z$. For some integers X and Y if $pX + aY \in N$ then this is either greater than q or equal to q.

Let $pX + aY = qs + r = (px_1 + ay_1)s + r$

$p(X - x_1 s) + a(Y - y_1 s) = r$

Now r is of the form $pX + aY$. If r would be positive, then it must be greater than or equal to q. But clearly $r < q$. Therefore, $r = 0$. Thus any integer which can be expressed in the form $pX + aY$ is divisible by q. Also $p = p.1 + a.0$. Thus p is divisible by q. But p is prime number and hence $q = 1$ or $q = p$. Since q divides $a = p.0 + a.1$, q cannot be equal to p. So $q = 1$. Thus

$$px_1 + ay_1 = 1$$

$$px_1 b + ay_1 b = b$$

Since p divides both $px_1 b$ and $ay_1 b$, we conclude that p divides b.

Proof of Theorem 2.1.2: Let $p_1^{a_1} p_2^{a_2} \dots p_k^{a_k} = c = q_1^{b_1} q_2^{b_2} \dots q_j^{b_j}$. Since $q1$ divides $p_1^{a_1} p_2^{a_2} \dots p_k^{a_k}$, q_1 must divide some p_1. Again, since q_1 and p_1 are prime numbers, $p_1 = q_1$. Similarly each p is equal to some q and each q is equal to some p. So $k = j$ and $p_1 = q_1, p_2 = q_2, \dots p_k = q_k$. If $a_1 < b_1$ then since,

$$p_2^{a_2} p_3^{a_3} \dots p_k^{a_k} = q_1^{a_1 - b_1} q_2^{b_2} \dots q_k^{b_k}$$

q_1 will divide at least one $p_2, p_3, p_4, \dots, p_k$. So $a_1 < b_1$ and similarly $b_1 < a_1$ are impossible. Hence $a_1 = b_1$ and similarly, $a_2 = b_2$, $a_3 = b_3$,... $a_k = b_k$ and this completes the proof.

Let a and b be two given positive integers.

Consider the set $\{ax + by{:}\, x$ and y are integers$\}$. Such a set is called a *Number Ring*. Let d be the smallest positive integer which belongs to this set. Thus there are integers x_1 and y_1 such that $d = ax_1 + by_1$ and if $ax + by$ is a positive integer then this is either greater than or equal to d.

We now prove that d divides each element of the number ring and if some integer n divides a and b both, then n divides d.

Proof: Let $a = dq + r$ where $0 \le r < d$.

$$r = a - dq = a - (ax_1 + by_1)q$$

$$= a(1 - qx_1) + b(-qy_1) = ax_2 + by_2$$

If r would be positive then it would be greater than or equal to d. But this being impossible $r = 0$. Thus d divides a and similarly d divides b.

If there is some integer n which divides a and b both then $a = nk$ and $b = nl$ for some integers k and l. So $d = ax_1 + by_1 = (kx_1 + by_1)u$ and hence $n|d$.

Note: d defined as above is said to be the *Highest Common Factor* (H.C.F) of a and b and is written as (a, b).

It is clear that if $d = (a, b)$ in order that the equation $ax + by = n$ will have integer solutions d must divide n. On the other hand if d divides n then we can find atleast one pair of integers (x, y) for which $ax + by = n$.

If $(a, b) = 1$ then a and b are said to be *co-prime* and in this case a and b do not have any common factor other than 1.

Similarly for given m positive integers $a_1, a_2, \cdots, a_m$ we can find a positive integer d such that d divides each a_k and if any other positive integer n divides each a_k then n must divide d. This d is called the H.C.F of $a_1, a_2, \cdots, a_m$ and written as $d = (a_1, a_2, \cdots, a_m)$. As before we can find m integers $x_1, x_2, \cdots, x_m$ for which,

$$d = a_1x_1 + a_2x_2 + \cdots + a_mx_m$$

We now prove this by induction on m. Firstly note that the result is true for m=2. If m> 2, then observe that $(a_1, a_2, \cdots a_m) = ((a_1, a_2, \cdots, a_{m-1}), a_m)$. For if $d_{m-1} = (a_1, a_2, \cdots, a_{m-1})$ and $d_m = (a_1, a_2, \cdots, a_m)$ then d_m/a_i $(i = 1, 2, \cdots, m)$. So $d_m\,|d_{m-1})$, but $d_m\,|a_m$. On the other hand if $c|d_{m-1})$ and $c|a_m$ then $c|d_m$. So $d_m = (d_{m-1}, a_m)$.

Now assume that if $d_r = (a_1, a_2, \cdots, a_r)$ then there are integers $y_1, y_2, \cdots, y_r$ for which $d_r = a_1y_1 + a_2y_2 + \cdots + a_ry_r$ holds. Then for two integers z_{r+1} and w_r we have,

$$\begin{aligned} d_{r+1} &= (d_r, a_{r+1}) = d_rw_r + a_{r+1}z_{r+1} \\ &= a_1(y_1w_r) + a_2(y_2w_r) + \cdots + a_r(y_rw_r) + a_{r+1}z_{r+1} \\ &= a_1z_1 + a_2z_2 + \cdots + a_rz_r + a_{r+1}z_{r+1} \end{aligned}$$

This completes the proof.

The H.C.F. of two natural numbers a and b is obtained by *Euclid's algorithm* which is described below.

Firstly if $a = b$ then $(a, b) = (a, a) = a$. Assume that $a \neq b$ and to be more specific let $b < a$. Divide a by b and let r_1 be the remainder. Then divide b by r_1 and let r_2 be the remainder. Now divide r_1 by r_2 and let r_3 be the remainder. Continuing in this way we get $r_4, r_5 \cdots$ such that $b > r_1 > r_2 > r_3 > \cdots r_k > r_{k+1} > \cdots$. Since $(r_1, r_2, \cdots)$ is a decreasing sequence of integers, there is some integer n such that $r_{n+1} = 0$.

The above fact can be expressed by the following equations,

$$a - q_1 b = r_1$$
$$b - q_2 r_1 = r_2$$
$$r_1 - q_3 r_2 = r_3$$
$$\dots\dots$$
$$r_{n-2} - q_n r_{n-1} = r_n$$
$$r_{n-1} - q_{n+1} r_n = 0$$

Put $r_n = h$. It is clear from the last equation that $h|r_{n-1})$. Since $h|r_n$ and $h|r_{n-1})$ it follows that $h|r_{n-2})$. Similarly it follows that $h|r_{n-3}, h|r_{n-4}, \cdots, h|r_2, h|r_1$. Thus $h|b$ and $h|a$.

On the other hand if $d|a$ and $d|b$ then d divides $r_1 = a - q_1 b$. Again since $d|b$ and $d|r_1$, d divides $r_2 = b - q_2 r_1$. Similarly it follows that d divides $r_3 = r_1 - q_3 r_2$. Continuing in this way finally we observe that d divides $h = r_n = r_{n-2} - q_n r_{n-1}$. Thus $(a, b) = h$ and finally we observe that,

$$\begin{aligned} h = r_n &= r_{n-2} - q_n r_{n-1} = r_{n-2} - q_n(r_{n-3} - q_{n-1} r_{n-2}) \\ &= (-q_n) r_{n-3} + (q_n q_{n-1} + 1) r_{n-2} \\ &= (-q_n) r_{n-3} + (q_n q_{n-1} + 1)(r_{n-4} - q_{n-2} r_{n-3}) \\ &= (q_n q_{n-1} + 1) r_{n-4} + (-q_n - q_{n-2} - q_n q_{n-1} q_{n-2}) r_{n-3} \\ &= \cdots = aP_n + bQ_n. \end{aligned}$$

For example, consider 741 and 1079. Then,

```
742)1079(1
     741
----------
338)741(2
    676
---------
65)338(5
   325
--------
13)65(5
   65
-------
    0
```

Hence $(741, 1079) = 13$. Further

$$\begin{aligned} 13 &= 338 - 325 = 338 - 5(65) = 338 - 5\{741 - 2(338)\} \\ &= 11(338) - 5(741) = 11(1079 - 741) - 5(741) \\ &= 1079(11) + 741(-16) = a_1 x_1 + a_2 x_2 \end{aligned}$$

where, $x_1 = 11$ and $x_2 = -16$.

We now discuss the Lowest Common Multiple (L.C.M) of two positive integers a and b.

Definition

If m is such an integer that (i) $a|m, b|m$ and (ii) if $a|c$, $b|c$ then $m|c$, then m is said to be the L.C.M. of a and b and is written as $m = [a, b]$. It is clear that there cannot be more than one integer satisfying the above two conditions. We now prove that $[a, b](a, b) = a.b$.

Proof: Let $(a, b) = d$, $a = a_1 d$ and, $b = b_1 d$ and $(a_1, b_1) = 1$. It now suffices to prove that $[a, b] = a_1 b_1 d = \frac{a.b}{d}$. Clearly $a|ab_1 = a_1 b_1 d$ and $b|ab_1 = a_1 b_1 d$.

Secondly if $a|c$ and $b|c$ then $c = c_1$, $a = c_1 a_1 d$, $c = c_1 b = c_1 b_1 d$. So $c_1 a_1 = c_1 b_1$. Since $(a_1, b_1) = 1$, $b_1 |c_1$. Thus take $c_1 = kb_1$. So $c = c_1 a_1 d = kb_1 a_1 d$, i.e. $a_1 b_1 d|c$, so $a_1 b_1 d = [a, b]$ and $ab = [a, b].(a, b)$.

Definition

Let $a_1, a_2, \cdots a_k$ be given positive integers. Let m be such an integer that,

(i) $a_i\ |m$ for $i = 1, 2, \cdots, k$
(ii) If $a_i\ |c$ for $i = 1, 2, \cdots, k$ then $m|c$.

Then m is said to be the L.C.M. of $a_1, a_2, \cdots a_k$ and is written as,

$$m = [a_1, a_2, \cdots a_k]$$

We will prove that if for $i = 1, 2, \cdots, k$, $a_i = p_1^{x_{i1}} p_2^{x_{i2}} \cdots p_r^{x_{ir}}$ (p_m being prime and $x_{ij} \geq 0$) and for $j = 1, 2, \cdots, r$, v_j is the greatest of all x_{ij} ($i = 1, 2, \cdots, k$) then,

$$[a_1, a_2, \cdots a_k] = p_1^{v_1} p_2^{v_2} \cdots p_r^{v_r}$$

Proof: Clearly for each i, $a_i = p_1^{v_1} p_2^{v_2} \cdots p_r^{v_r}$. If for $i = 1, 2, \cdots, k$, $a_i\ |c$ and if for $i = s, x_{i1} = v_1$, then from a_s/c we get $p_1^{v_1}\ |c$. Similarly, each one of $p_2^{v_2} p_3^{v_3} \cdots p_r^{v_r}$ divides c and hence $p_1^{v_1} p_2^{v_2} \cdots p_r^{v_r}\ |c$.

In this section we discuss some properties of prime numbers.

Theorem 2.1.4 (Euclid's Second Theorem)

There are infinitely many prime numbers.

Proof: Let p be a prime number. It is enough to show that there is a prime number greater than p. Let

$$q = (2.3, 5.7, 7.11 \cdots p) + 1$$

Now q is not divisible by any one of $2, 3, 5, 7, 11, \cdots, p$. Therefore, every prime factor of q is greater than p, i.e. either q is divisible by a prime number great than p or q itself is prime. This completes the proof.

It may be noted here that every prime number greater than 2 is an odd integer, so it can be expressed by $4n + 1$ or $4n + 3$.

There are infinitely many prime numbers which can be expressed in the form $4n + 3$.

Proof: Firstly, $3 = 4.0 + 3$ is a prime number. If $Q = (2^2.3.5.7.11 \cdots p) + 1$, then Q can be expressed in the form $(4n + 3)$. If none of $2, 3, \cdots, p$ divide Q, so every prime factor of Q is greater than p. On the other hand Q cannot be expressed as a product of prime numbers of the form $(4n + 1)$, since $(4n_1 + 1)(4n_2 + 1) \cdots (4n_k + 1) = (4n + 1)$. Thus Q will be divisible by a prime number greater than p and expressible in the form$(4n + 3)$. Therefore given a prime p there always exists a prime number greater than p and expressible in the form $(4n + 3)$ and this proves the assertion.

2.2 CONGRUENCE

Definition

Let m be a positive integer and $m|(x - a)$. Then x is said to be congruent to $a \bmod m$ and written as $x \equiv a(mod\, m)$. Then a is said to be a residue of $x \bmod m$. Again if $0 \leq a \leq m - 1$, then a is said to be the least residue of $x \bmod m$.

For a given residue, the set of all integers congruent to this $modm$ is called the *class of residues* $modm$. The difference of any two elements of this class is divisible by m.

Example 2.2.1

Let $m = 4, a = 3$. Then if $x \equiv 3\ (mod\ 4), 4|(x - 3)$, i.e., $x - 3 = 4q$ or $x = 4q + 3$. By taking different integer values of q we obtain the set of values of x:

$$\{\cdots, -17, -13, -9, -5, -1, 3, 7, 11, 15, 19, \cdots\}$$

We denote this set by $\bar{3}$. Similarly,

$$\bar{0} = \{\cdots, -4, 0, 4, 8, \cdots\}$$
$$\bar{1} = \{\cdots, -7, -3, 1, 5, 9, \cdots\}$$
$$\bar{2} = \{\cdots, -6, -2, 2, 6, 10, \cdots\}$$

Similarly, if $m = 5$ we get,

$$\bar{0} = \{\cdots, -15, -10, -5, 0, 5, 10, \cdots\}$$
$$\bar{1} = \{\cdots, -14, -9, -4, 1, 6, 11, \cdots\}$$
$$\bar{2} = \{\cdots, -13, -8, -3, 2, 7, 12, \cdots\}$$
$$\bar{3} = \{\cdots, -12, -7, -2, 3, 8, 13, \cdots\}$$
$$\bar{4} = \{\cdots, -11, -6, -1, 4, 9, 14, \cdots\}$$

It is clear that for a given m, the m number of residue classes are as follows:

$$0^{-} = \{\cdots, -3m, -2m, -m, 0, m, 2m, 3m, \cdots\}$$
$$1^{-} = \{\cdots, -3m + 1, -2m + 1, -m + 1, 1, m + 1, 2m + 1, 3m + 1, \cdots\}$$
$$2^{-} = \{\cdots, -3m + 2, -2m + 2, -m + 2, 2, m + 2, 2m + 2, 3m + 2, \cdots\}$$

$\cdots\cdots\cdots$

$$r^{-} = \{\cdots, -3m + r, -2m + r, -m + r, r, m + r, 2m + r, 3m + r, \cdots\}$$
$$(m - 1)^{-} = \{\cdots, -2m - 1, -m - 1, -1, m - 1, 2m - 1, 3m - 1, \cdots\}$$

The set consisting of $0^{-}, 1^{-}, 2^{-}, \cdots, (m - 1)^{-}$ or the set formed by choosing one representative from each of the above sets is called a complete system of incongruent residues.

Some properties of congruence: It is clear that for a given m,

(i) $a \equiv a$ (reflexivity),
(ii) $a \equiv b \Rightarrow b \equiv a$ (symmetry),
(iv) $a \equiv b \Rightarrow b \equiv c$ (transitivity),
(v) $a \equiv b$ and k is any integer $\Rightarrow ka = kb$,
(vi) $a \equiv b$ and $a' \equiv b' \Rightarrow a + a' = b + b'$ and $aa' = bb'$,

Further for any positive integer n, $a^n \equiv b^n$.

Note that for a given m, the relation. "Congruence mod m" is an equivalence relation and hence defines a partition in the set of all integers.

EXAMPLES OF PROOF BY INDUCTION

Prove the following statements by induction on n:

1. $\sum_{i=1}^{n} i = \frac{1}{2} n(n+1), (n \geq 1)$
2. $\sum_{i=1}^{n} i^2 = \frac{1}{6} n(n+1)(2n+1), \ (n \geq 1)$
3. $\sum_{i=1}^{n} i^3 = \left[\frac{1}{2} n(n+1)\right]^2, (n \geq 1)$
4. $(A_1 \cup A_2 \cdots \cup A_n)' = A_1' \cap A_2' \cdots \cap A_n', \ (n \geq 1)$
5. $(A_1 \cap A_2 \cdots \cap A_n)' = A_1' \cup A_2' \cdots \cup A_n', \ (n \geq 1)$
6. For a, $b \in R$ and $n \geq 1$, $(a + b)^n = a^n + {}^nC_1 \ a^{n-1} \ b + {}^nC_2$ $a^{n-2} \ b^2 + \cdots + b^n$.

We prove only 1, 4 and 6; others are left to the reader as exercises.

***Proof of* (1):** For $n = 1$, the statement is true, because in this case

$$\sum_{i=1}^{1} i = \frac{1}{2} 1(1+1) = \frac{1}{2} . 1.2 = 1.$$

The induction hypothesis is that for all $k \geq 1$,

$$\sum_{i=1}^{k} i = \frac{1}{2} k(k+1)$$

Now,

$$\sum_{i=1}^{k+1} i = \left(\sum_{i=1}^{k} i\right) + (k+1)$$

$$\left(\frac{1}{2} k(k+1)\right) + (k+1) = \frac{1}{2}(k+1)(k+2)$$

which proves the statement for $n = k + 1$. This completes the proof.

Proof of (4): For $n = 1$, the statement is trivially true. The induction hypothesis is that for all $k \geq 1$,

$$(A_1 \cup A_2 \cdots \cup A_k)' = A_1' \cap A_2' \cdots \cap A_k'$$

Using De-Morgan's laws we have,

$$(A_1 \cup A_2 \cdots \cup A_{k=1})' = [(A_1 \cup A_2 \cdots \cup A_k) \cup A_{k+1}]'$$
$$= (A_1 \cup A_2 \cdots \cup A_k)' \cap A_{k+1}'$$
$$= A_1' \cap A_2' \cdots \cap A_k' \cap A_{k+1}'$$

which proves the statement for $n = k + 1$ and this completes the proof.

Proof of (6): For $n = 1$, the statement is clearly true. The induction hypothesis is that for all $k \geq 1$,

$$(a+b)^k = a^k + {}^kC_1 a^{k-1}b + \cdots + b^k.$$

Now,

$$(a+b)^{k+1} = (a+b)^k \ (a+b)$$
$$= \left(a^k + {}^kC_1 a^{k-1}b + \cdots + b^k\right) \ (a+b)$$
$$= a^{k+1} + \left({}^kC_1 + 1\right)a^k b + \cdots + b^{k+1}$$
$$= a^{k+1} + {}^{k+1}C_1 \ a^k b + \cdots + b^{k+1}$$

which proves the statement for $n = k + 1$ and this completes the proof.

EXERCISES

1. If for $i \neq j$, $(a_i, a_j) = 1$ then prove that $[a_1, a_2, \cdots, a_k] = a_1 a_2 \cdots a_k$.
2. Prove that, $[a_1, a_2, \cdots, a_k] = [[a_1 a_2 \cdots, a_{k-1}], a_k]$
3. If $(a, b) = 1$ and c/a then show that $(c, b) = 1$.
4. Prove that, $(ma, mb) = m(a, b)$
5. If $(a, b) = 1$ then show that, $(ac, b) = (c, b)$
6. $\sum_{i=1}^{n} i = \frac{1}{2}n(n+1)(n \geq 1)$.
7. $\sum_{i=1}^{n} i^2 = \frac{1}{6}n(n+1)(2n+1)\ (n \geq 1.)$
8. $\sum_{8.}^{n} i^3 = \left[\frac{1}{2}n(n+1)\right]^2 (n \geq 1)$.

$i=1$

9. $1.2 + 2.3 + \cdots + n(n+1) = \frac{n(n+1)(n+2)}{3}$.
10. $1^2 + 3^2 + 5^2 + \cdots + (2n-1)^2 = \frac{n(2n+1)(2n-1)}{3}$.
11. $2 + 5 + 8 + \cdots + (3n-1) = \frac{3(3n+1)}{2}$.
12. $1 + 5 + 9 + \cdots + (4n-3) = n(2n-1)$.
13. $2 + 2^2 + \cdots + 2^n = 2^{n+1} - 2$.
14. $(a+b)^n = a^n + \binom{n}{1} a^{n-1}b + \binom{n}{2} a^{n-2}b^2 + \cdots \ldots + b^n$.
15. $2^n > n^3$ for $n \geq 10$.
16. $1 + 2^n < 3^n$for $n \geq 2$.
17. $2^n > n$ for $n > 1$.
18. Any positive integer$n > 2$ is either a prime or a product of primes.
19. $n^3 + 2n$ is divisible by 3 for $n \geq 1$.
20. $n^4 - 4n^2$is divisible by 3 for $n \geq 2$.

CHAPTER 3

Group Theory

3.1 DEFINITIONS AND EXAMPLES

Definitions

Let G be a non empty set. Let there be a mapping from $G \times G$ into G called a *binary composition*, defined by,

$$(a, b) \rightarrow ab$$

Then G along with this binary composition is called a group if it *satisfies* the following properties:

1. For $a, b \in G$ we have $ab \in G$ (closure property),
2. For $a, b, c \in G$ we must have
   $a(bc) = (ab)c$ (associative law)
3. There exists an element $e \in G$, called the identity, such that for every $a \in G, ae = ea = a$ (the existence of an identity),
4. For every $a \in G$, there exists an element $a^{-1} \in G$ called the inverse of a in G such that,

   $aa^{-1} = a^{-1}a = e$ (the existence of inverse)

   The property (i) is a superfluous statement, it only tells again that the mapping $(a, b) \rightarrow ab$ is from $G \times G$ into G, i.e. it is a binary composition. G with property (i) only is called a groupoid, with (i) and (ii) is called a semigroup and with (i), (ii) and (iii) only is called a *monoid.*

   A group G is called *abelian* or *commutative* if it has the following property.
5. For $a, b, \in G, ab = ba$ (commutative law). Otherwise a group G is called *non-abelian.*

A group G is called finite if it is a finite set, otherwise it is called an *infinite group.* The number of elements of a group G is called the *order* of G and is denoted by $o(G)$. This number is most interesting, useful and meaningful when G is a finite group.

Remark

In the above definitions we have used multiplicative notation for the binary composition. If one uses additive notation the definition of a group reads as follows.

A non-empty set G along with a mapping $+: G \times G \rightarrow G$ defined by,

$$(a, b) \rightarrow a + b$$

Is called a group if:

1. $a, b \in G$ implies $a + b \in G$
2. $a, b, c \in G$ implies $a + (b + c) = (a + b) + c$
3. For every $a \in G$, there exists an element $0 \in G$ such that,
   $a + 0 = 0 + a = a$
4. For every $a \in G$, there exists an element $(-a) \in G$ such that,
   $a + (-a) = (-a) + a = 0$

   However while proving the general results we shall use the multiplicative notation for the binary composition. Before we go further we shall first give some examples of groups leaving the verification for the reader as exercise.

3.1.1 Examples of Groups

Examples 3.1

1. With respect to usual addition of numbers the following sets form groups:
   (a) The singleton set consisting only of the number 0.
   (b) The set **Z** of all integers. The set of all nonnegative integers does not form a group.
   (c) The set of all even integers including 0. Observe that the set of all odd integers is not a group.
   (d) The set **Q** of all rational numbers.
   (e) The set **R** of all real numbers.
   (f) The set **C** of all complex numbers.
   (g) The set of all complex numbers, whose real and imaginary parts are integers.
2. The following sets of numbers form groups with respect to usual multiplication of numbers.
   (a) The singleton set consisting only of the number 1,
   (b) The set $\{1, -1\}$.

(c) $\{1, -1, i, -i\}$.
(d) The set **Z** of all integers is not a group.
(e) The set of all positive rational numbers (excluding 0). The set of all nonzero rationals. Observe that the set of all negative rationals is not a group.
(f) The set of all positive real numbers. Note that the negative real numbers do not form a group.
(g) The set of all nonzero real numbers. The set **R** of all real numbers does not form a group.
(i) The set of all nonzero complex numbers. Observe that the set **C** of all complex numbers does not form a group.
(j) The unit circle $\{z \in C: |z| = 1\}$. This is called the circle group.
(k) The set $\{z \in C: z^n = 1\}$ where n is a fixed but arbitrary positive integer.

3. The set **R**" of all n-tuples of real numbers, i.e.,

   R" $= \{x = (x_1, \dots, x_n): x_i \in \mathbf{R}, i = 1, \dots, n\}$

   Is a group with respect to coordinate wise addition defined by,

   $x + y = (x_1 + y_1, \dots, x_n + y_n)$

   Where $x = (x_1, \dots, x_n)$ and$y = (y_1, \dots, y_n)$. Similarly the set **C**" of all n-tuples of complex numbers form a group with respect to coordinate wise addition.

4. The family of all subsets of a set U is a group with respect to the formation of symmetric difference of two sets A and B defined by,

   $A \Delta B = (A - B) \cup (B - A)$

   It should be noted that if $U \neq \varphi$, then the above family of sets does not constitute a group with respect to the formation of either unions of intersections.

5. Let m be a positive integer and let $\mathbf{Z}_m$, be the set of all nonnegative integers less than m, i.e. $\mathbf{Z}_m = \{0, 1, \dots, m-1\}$. If a and b are in $\mathbf{Z}_m$ define their sum $a + b$ to be the remainder obtained when their usual sum is divided by m. Then $\mathbf{Z}_m$ is a group with respect to the operation defined above.

6. The set of all 1–1 mappings of a nonempty set X onto itself is a group with respect to the multiplication (composition) of mappings defined as follows: if f and g are two such mappings and x is an arbitrary element of X then $(fg)(x) = f(g(x))$.

7. As a special case of the above example let $X = \{1, 2, \dots, n\}$. This case is of particular importance. A 1–1 mapping of this set X onto itself is called a permutation and can be written in the convenient form,

$$\begin{pmatrix} 1 & 2\ldots & n \\ i_1 & i_2\ldots & i_n \end{pmatrix}$$

where below each of the integers $1, 2, \ldots, n$ is placed its image under the mapping.

For example, if $n = 4$ two such mappings p and q can be conveniently written in the form,

$$p = \begin{pmatrix} 1 & 2 & 3 & 4 \\ 3 & 1 & 4 & 2 \end{pmatrix}, \quad q = \begin{pmatrix} 1 & 2 & 3 & 4 \\ 4 & 1 & 3 & 2 \end{pmatrix}$$

We define the product pq of p and q by,

$$\begin{pmatrix} 1 & 2 & 3 & 4 \\ 2 & 3 & 4 & 1 \end{pmatrix}$$

The set of all permutations of n elements is denoted by S_n and this forms a group under the product of permutations. This group is called the symmetric group of degree n.

8. Consider the set of all $m \times n$ (with m and n fixed positive integers) matrices with real or complex entries. If $A = (a_{jk})$ and $B = (b_{jk})$ are two such matrices then we define the sum $A + B$ to be the matrix $C = (c_{jk})$ with elements,

   $$c_{jk} = a_{jk} + b_{jk}, j = 1, \ldots, m, k = 1, \ldots, n$$

   The above set forms a group with respect to the operation of matrix addition defined as above. In particular the set of all square $n \times n$ matrices forms a group under matrix addition, but this set does not form a group under matrix multiplication. Any n-rowed square matrix A need not have an inverse A^{-1}such that,

   $$AA^{-1} = A^{-1}A = I$$

   where I is the n-rowed identity (unit) matrix. This holds if and only if A is nonsingular, that is, if and only if $detA \neq 0$ where $detA$ denotes the determinant of A.

9. The set of all n^{th} roots of 1 is a group with respect to multiplication. If A denotes the set of all n^{th} roots of 1 then,

   $$A = \{e^{2\pi ir/n} : r = 0, 1, 2, \ldots, (n-1)\}$$

   If $a = e^{2\pi_i r_1/n}$, $0 \leq r_1 \leq n-1$ and $b = e^{2\pi_i r_2/n}, 0 \leq r_2 \leq n-1$, then $ab = e^{2\pi_i (r_1+r_2)/n}$. If $r_1 + r_2 \leq n-1$, then $ab \in A$. If $r_1 + r_2 \geq n$ then assume that $r_1 + r_2 = tn + k, 0 \leq k \leq n-1$ and then,

   $$ab = e^{2\pi i(tn+k)/n} = e^{2\pi it} . e^{2\pi ik/n}$$

   $$= e^{2\pi ik/n} \in A \text{ since } 0 \leq k \leq n-1$$

The verification of other axioms of group are left to the reader. It should be noted that this group is abelian, for

$$e^{2\pi i r_1/n}.e^{2\pi i r_2/n} = e^{2\pi i(r_1+r_2)/n}$$
$$= e^{2\pi i(r_2+r_1)/n} = e^{2\pi i r_2/n}.e^{2\pi i r_1/n}$$

3.1.2 Worked Out Examples

Example 3.2

1. Show that if every element of the group G is its own inverse, then G is abelian.
   Let $a, b \in G$. Then $a^{-1}, b^{-1} \in G$, $ab \in G$ and $(ab)^{-1} \in G$. But $a = a^{-1}, b = b^{-1}$. Now $ab = (ab)^{-1} = b^{-1}a^{-1} = ba$.
   Hence, G is abelian.
   For any positive integer n we write $a^n = aaa \dots a$ (n times).
2. If $a^2 = e$ for every element a in a group G, then show that G is abelian,
   We have $a^2 = aa = e.$ But $aa^{-1} = e.$
   Therefore, $aa = e = aa^{-1}$ and so $a = a^{-1}$ for every $a \in G$. Now it follows from the previous example that G is abelian.
3. If a group G has four elements, then show that it must be abelian.
   Let $G = \{e, a, b, c\}$ where e is the unit element in G.
   If each element is its own inverse, then G is abelian. If not, assume that $a^{-1} = b$ and $c^{-1} = c$.
   That is , $ab = ba = e$ and $cc = e$. Now

   $ac \neq e$ since $a^{-1} = c$

   $ac \neq a$ since $c \neq e$

   $ac \neq c$ since $a \neq e$

   Therefore $ac = b$. Similarly it can be shown that $ca = b$. Thus $ac = ca$.

   $bc \neq e$ since $b^{-1} \neq c$

   $bc \neq b$ since $c \neq c$

   $bc \neq c$ since $b \neq e$

   Therefore $bc = a$ and similarly $cb = a$ so that $bc = cb$. Thus G is abelian. The binary composition in G which we call multiplication, is given by the following table,

	e	a	b	c
e	e	a	b	c
a	a	c	e	b
b	b	e	c	a
c	c	b	a	e

4. If G is a group such that $(ab)^2 = a^2b^2$ for all $a, b \in G$ show that G must be abelian.

   For all $a, b \in G$ we have

   $(ab)^2 = a^2b^2$

   i.e. $(ab)(ab) = (aa)(bb)$

   i.e. $a(ba)b = a(ab)b$`

   i.e. $a(ba) = a(ab)$ (by right cancellation)

   i.e. $ba = ab$ (by left cancellation)

   Therefore G is abelian.

Definition

If G is a group and $a \in G$ then the order of a is defined to be the least positive integer n such that$a^n = e$. If there exists no such positive integer then a is said to be of infinite order. We write $o(a)$ to denote the order of a.

Example 3.3

1. In any group $G, o(e) = 1$, where e denotes the identity element in G.
2. In the multiplicative group $\{1, -1, i, -i\}, o(-1) - 2, o(i) - 4$.
3. In the group

   $\{..., a^{-3}, a^{-2}, a^{-1}, a^0 = e, a, a^2, a^3, ...\}$

   the order of any element is either 0 or ∞.
4. If w denotes a cube root of 1 then in the group.

   $\{1, \omega, \omega^2\}, o(1) = 1, o(\omega) = o(\omega^2) = 3$

Theorem 3.1

1. Let G be a group and $a \in G$. Then for any positive integer $k, o(a^k) \leq o(a)$.
2. $o(a^{-1}) = o(a)$.

Proof: (i) Let $o(a) = n$ and $o(a^k) = m$. Therefore $a^n = e$ and $(a^k)^m = e$. So $(a^k)^n = (a^n)^k = e^k = e$. But m is the smallest positive integer for which $(a^k)^m = e$. So $m \leq n$.

Similarly,

$$a^m = ((a^{-1})^{-1})^m = (a^{-1})^{-m} = e^{-1} = e \text{ and so } n \leq m.$$

Therefore, $m = n$ and this completes the proof.

Theorem 3.2

The order of every element of a finite group G is finite.

Proof: Let $o(G) = n$ and let $a \in G$ be arbitrary. The elements $e, a, a^2, \dots$, a^n are $(n+1)$ in number and $o(G) = n$ and hence they are not all distinct. In other words, there exist r, s with $1 \leq r \leq s \leq n$ such that $a^r = a^s$. Hence $e = a^r a^{-r} = a^s a^{-r} = a^{s-r} = a^k$ where $k = s - r, 1 \leq k \leq n$.

Therefore $o(a)$ is finite and since a is arbitrary, this completes the proof.

Definition

Let G be a group. If for some $a \in G$ each element of G is precisely of the form a^n for some $n \in Z$ then G is said to be a cyclic group and a is said to be generator of G. Note that if G is a cyclic group and a is its generator, then,

$$G = \{a^n : n \in Z\} = (a)$$

Example 3.4

(a) Under addition,

$$Z = \{\dots, -2, -1, 0, 1, 2, \dots\}$$

is a cyclic group with 1 as the generator. Also under addition,

$$\{\dots, -4, -2, 0, 2, 4, \dots\}$$

is a cyclic group with 2 as generator.

(b) $\{\dots, a^{-2}, a^{-1}, a^{-0} = e, a^1, a^2, \dots\}$ is a cyclic group with generator a.

(c) Under multiplication $\{1, -1, i, -i\}$ is a cyclic group with i as a generator, $-i$, being a generator also.

Theorem 3.3

1. Every cyclic group is abelian.
2. The order of a cyclic group is same as that of its generator.

Proof:

1. Let $G = (a) = \{e, a, a^2, a^3, ...\}$ be a cyclic group with a as the generator. Then for any $m, n \in Z, a^m, a^n \in G$ and $a^m a^n = a^{m+n} = a^{n+m} = a^n a^m$ and hence G is abelian. In the additive notation we write,

   $G = \{1, a, 2a, 3a, ...\}$

   In this case,

   $ma + na = (m+n)a = (n+m)a = na + ma$ and thus G is abelian.
2. Let G be a cyclic group with generator $a \in G$. Let $o(a) = n$, i.e.

   $a^n = e$ and $a^m \neq$ e for $0 < m < n$

   if $m > n$, let $m = qn + r, 0 \leq r \leq n.$

   So,

   $$a^m = a^{qn+r} = a^{qn}.a^r = (a^n)^q.a^r$$
   $$= e^q.a^r = e.a^r = a^r$$

   Therefore there are exactly n elements in the group and they are,

   $\{e, a, a^2, ..., a^{n-1}\}$

   This completes the proof.

3.2 SUBGROUPS

Definition

Let G be a group and let H be a non empty subset of $G. H$ is said to be a subgroup of G if under the composition in G, H itself forms a group. If H is a subgroup of G then we denote this by $H < G$.

3.2.1 Examples

Example 3.5

1. The singleton set $\{e\}$ and the whole group G, where e is the identity element of G are subgroups of any group G. These are called trivial subgroups or improper subgroups. Any other subgroup is called a proper or nontrivial subgroup.
2. The group of all even integers is a subgroup of the group of all integers under addition. But the set of all odd integers is not a subgroup of all integers under addition. Similarly the set of all positive integers is not a subgroup of all integers under addition.

3. The group of all integers is a subgroup of all rational numbers under addition.
4. The group of positive rational numbers under multiplication is a subgroup of nonzero real numbers under multiplication.
5. $\{1, -1, i, -i\}$ is a subgroup of the group of all nonzero complex numbers under multiplication.
6. $\{1, -1\}$ is a subgroup of $\{1, -1, i, -i\}$ under multiplication.

Theorem 3.4

Let G be a group and let $\phi \neq H \subset G$.

Then $H < G$ if and only if:

(i) $a, b \in H$ if implies that $ab \in H$.
(ii) $a, b \in H$ implies that $a^{-1} \in H$.

Proof: If $H < G$ then it clearly follows from the definition that (i) and (ii) hold.

Conversely, suppose that $\phi \neq H \subset G$ for which (i) and (ii) hold. Since associative law holds in G and $H \subset G$ it also holds in H. So in order to prove that $H < G$ it is now enough to show that $e \in H$. For this, let $a \in H$. By (ii) it follows that $a^{-1} \in H$. Now by (i) we get $e = aa^{-1} \in H$ and this completes the proof.

Theorem 3.5

Let G be a group and let $\phi \neq H \subset G$. Then $H < G$ if and only if $ab^{-1} \in H$ whenever $ab \in H$.

Proof: Let $H < G$ and let $a, b \in H$. Since $b \in H$ we have $b^{-1} \in H$. Again since $a \in H$ we have $ab^{-1} \in H$. Conversely, let $\phi \neq H \subset G$ for which $ab^{-1} \in H$ whenever $a, b \in H$. Let $a \in H$. Then $aa^{-1} = e \in H$. Now for each $b \in H$ we have $eb^{-1} = b^{-1} \in H$. Let $a, b \in H$. Then $b^{-1} \in H$ and hence $a(b^{-1})^{-1} = ab \in H$. The associative law clearly holds and hence $H < G$.

Theorem 3.6

If H is a finite nonempty subset of a group G and H is closed under the composition, then H is a subgroup of G.

Proof: We need only to show that whenever $a \in H$ then $a^{-1} \in H$. Let $a \in H$. Then since H is closed under the composition we have $aa = a^2 \in H, a^2a =$

$a^3 \in H, a^3a = a^4 \in H, \dots, a^m \in H, \dots$ Thus the infinite set of elements $a, a^2, a^3, \dots, a^m, \dots$ must lie in H. But since H is a finite set, there must be some repetitions in the above set of elements, i.e. for some integers r and s with $r > s > 0$ we must have $a^r = a^s$. Therefore $a^{r-s} = a^{s-s} = e \in H$. Since $r - s - 1 \geq 0, a^{r-s-1} \in H$. Further since $aa^{r-s-1} = a^{r-s} = e$, we have $a^{-1} = a^{r-s-1}$. Thus $a^{-1} \in H$. This completes the proof of the theorem.

Theorem 3.7

If H_1 and H_2 are subgroups of a group G then $H_1 \cap H_2 < G$.

Proof: Since H_1 and H_2 are subgroups of a group $G, e \in H_1 \cap H_2$ and thus $H_1 \cap H_2 \neq \varphi$. Let $a, b \in H_1 \cap H_2$. Then $a, b \in H_1$ and since $H_1 < G. ab^{-1} \in H_1$. Similarly $ab^{-1} \in H_2$. Thus $ab^{-1} \in H_1 \cap H_2$ and shows that $H_1 \cap H_2 < G$.

Theorem 3.8

If H_1 and H_2 are subgroups of a group G, then $H_1 \cup H_2 < G$ if and only if:

$$H_1 \subset H_2 \; or \; H_2 \subset H_1.$$

Proof: If H_1 and H_2 are subgroups of a group G and if either $H_1 \subset H_2$ or $H_2 \subset H_1$ then it is clear that $H_1 \cup H_2 < G$.

Conversely, let $H_1 \cup H_2 < G$. We have shown that either $H_1 \subset H_2$ or $H_2 \subset H_1$. Assume, to the contrary, that neither holds. Then it is possible to choose elements a and b such that $a \in H_1 - H_2$ and $b \in H_2 - H_1$.

In other words $a \in H_1, a \notin H_2, b \in H_2, b \notin H_1$ but $a, b \in H_1 \cup H_2$. Since $H_1 \cup H_2$ is a subgroup and $a, b \in H_1 \cup H_2$ we must have $ab \in H_1 \cup H_2$. Therefore either $ab \in H_1$ or $ab \in H_2$. Let $ab \in H_1$. Since $a \in H_1$ we have $a^{-1} \in H_1$ and

$$b = a^{-1}(ab) \in H_1$$

which contradicts the fact that $b \notin H_1$. On the other hand if $ab \in H_2$. Since $b \in H_2$ and thus $b^{-1} \in H_2$ we must have,

$$a = (ab)b^{-1} \in H_2$$

But this is impossible. This contradiction completes the proof.

Definition

If H and K are any two subsets of a group G then we define,

$$HK = \{x \in G : x = hk, h \in H, k \in K\}$$

In the following theorem we shall assume that H and K are any two subgroups of a group G.

Theorem 3.9

HK is a subgroup of G if and only if $HK = KH$. In particular if G is abelian, then HK is a subgroup of G.

Proof: Suppose that $HK = KH$. In other words if $h \in H$ and $k \in K$ then $hk = k_1 h_1$ for some $k_1 \in K$ and $h_1 \in H$. We shall show that HK is a subgroup of G. First let $x = hk \in HK$ and $y = h'k' \in HK$. Then $xy = hkh'k'$. Now since $kh' \in KH = HK$ we have $kh' = h_2 k_2$ for some $h_2 \in H$ and $k_2 \in K$. Hence $xy = h(kh')k' = h(h_2k_2)k' = (hh_2)(k_2k') \in HK$. Further $x^{-1} = (hk)^{-1}h^{-1} \in KH = HK$ and thus $x^{-1} \in HK$ for each $x \in HK$. Therefore HK is a subgroup of G.

Conversely, let HK be a subgroup of G. Then for any,

$$h \in H \text{ and } k \in K, h^{-1} \in H. k^{-1} \in K \text{ and } so\ h^{-1}k^{-1} \in HK.$$

Therefore $kh = (h^{-1}k^{-1})^{-1} \in HK$.

Thus $KH \subset HK$. Now for any element $x \in HK, x^{-1} = hk \in HK$ so $x = (x^{-1})^{-1} = hk^{-1} = k^{-1}h^{-1} \in KH$ showing that $HK \subset KH$ and this completes the proof. The particular case follows from the fact that if G is abelian then $HK = KH$.

If H and K are finite subgroups of a group G, then although HK need not be a sub-group, it is clearly a finite set. Therefore it is a natural and a meaningful question to ask: How many distinct elements are there in the subset HK. If $o(HK)$denotes the number of distinct elements in the set HK then we have.

Theorem 3.10

If H and K are finite subgroups of a group G of orders $o(H)$ and $o(K)$ respectively, then

$$o(HK) = \frac{o(H)o(K)}{o(H \cap K)}$$

In particular if $H \cap K = \{e\}$ then $o(HK) = o(H)o(K)$.

Proof: Let $h \in H$ and $k \in K$ so that $hk \in HK$. If $h_1 \in H \cap K$ then $hk = (hh_1)({h_1}^{-1}k)$ where $hh_1 \in H$ since $h \in H$ and $h_1 \in H \cap K \subset H$ and ${h_1}^{-1}k \in K$ since ${h_1}^{-1} \in H \cap K \subset K$ and $k \in K$. Thus the element hk is duplicated atleast $o(H \cap K)$ times. However, if $hk = h_2k_2$, then $h^{-1}h_2 =$

$k(k_2)^{-1} = u$ and $u \in H \cap K$ and so $h_2 = hu, k_2 = u^{-1}k$ thus all the duplications are in the form $hk = (hh_1)(h_1{}^{-1}k)$. Therefore the element hk appears in the list of HK exactly $o(H \cap K)$ times. But the total number in the listing of HK is $o(H)o(K)$. Therefore the number of distinct elements in $HK, o(HK)$ is given by,

$$\frac{o(H)o(K)}{o(H \cap K)}$$

We shall now introduce two subgroups of a group G which are very useful and important.

Definition

If G is a group and $a \in G$ the normalizer of a in G is the set,

$$N(a) = \{x \in G : xa = ax\}$$

Theorem 3.11

$N(a)$ is a subgroup of G.

Proof: Suppose that $x, y \in N(a)$. Then $xa = ax$ and $ya = ay$. Therefore $(xy)a = x(ya) = x(ay) = (xa)y = (ax)y = a(xy)$ and so $xy \in N(a)$.

Now since $ax = xa$ it follows that,

$$x^{-1}a = x^{-1}(ax)x^{-1} = x^{-1}(xa)x^{-1} = ax^{-1}$$

So that $x^{-1} \in N(a)$ and this completes the proof.

Definition

If G is a group then the centre of G is defined by,

$$\text{cent } G = \{c \in G : xc = cx \text{ for all } x \in G\}$$

Theorem 3.12

Cent G is a subgroup of G.

Proof: First of all observe that cent $G \neq \varphi$ since $e \in cent\ G$. Let $a, b \in$ cent G. Then $xa = ax$ and $xb = bx$ for all $x \in G$. Therefore $b^{-1}(xb)b^{-1} = b^{-1}(bx)b^{-1}$ or $b^{-1}x = xb^{-1}$

Now for any $x \in G, (ab^{-1})x = a(b^{-1}x) = a(xb^{-1}) = (ax)b^{-1} = (xa)b^{-1}$ $= x(ab^{-1})$.

Therefore $ab^{-1} \in cent\ G$ and this proves that $cent\ G$ is a subgroup of G.

Definition

Let G be a group and H be a subgroup of G. Let $a, b \in G. a$ is said to be congruent to $b\ mod\ H$ written as $a \equiv b \text{ (mod } H) \ if\ ab^{-1} \in H$.

Theorem 3.13

The relation $a \equiv b \text{ (mod } H)$ is an equivalence relation.

Proof: We have to show that for $a, b, c \in G$:

(i) $a \equiv a \text{ (mod } H)$ (reflexivity)
(ii) $a \equiv b \text{ (mod } H)$ implies $b \equiv a \text{ (mod } H)$ (symmetry)
(iii) $a \equiv b \text{ (mod } H), b \equiv c \text{ (mod } H)$ implies $a \equiv c \text{ (mod } H)$ (transitivity).

Since H is a subgroup we have $e \in H$. But for any $a \in G, e = aa^{-1}$. Hence $aa^{-1} \in H$ and this proves the reflexivity. For the symmetry, let $a \equiv b \text{ (mod } H)$. Then $ab^{-1} \in H$. Since H is a subgroup.

$$(ab^{-1})^{-1} = (b^{-1})^{-1}a^{-1} = ba^{-1} \in H$$

Therefore $b \equiv a \text{ (mod } H)$ and this establishes (ii). Now let $a \equiv b \text{ (mod } H)$ and $b \equiv c \text{ (mod } H)$.

Then $ab^{-1} \in H$ and $bc^{-1} \in H$. Now,

$$ac^{-1} = a(b^{-1}b)c^{-1} = (ab^{-1})(bc^{-1}) \in H$$

Hence $a \equiv c\ (mod\ H)$ and this completes the proof.

Example 3.6

Let G be the group Z of all integers under addition. Let n be a fixed but arbitrary integer and let $H = H_n$ be the subgroup of Z consisting of all multiples of n. Then for $a, b \in G$ the relation $a \equiv b\ (mod\ H)$, i.e, $ab^{-1} \in H$ in this particular case under additive notation reads $a \equiv b \text{ (mod } n)$ if $a - b$ is a multiple of n.

This is the usual congruence $mod\ n$ discussed in Number Theory. Thus the relation defined above is the natural generalization of a familiar group to any arbitrary group and its subgroup.

Wc shall now elaborate the usual congruence relation by example.

Let n be a positive integer and $n|(x - a)$. Then x is congruent to $a\ mod\ n$ and written as $x \equiv a\ (mod\ n)$. On the other hand if $n \nmid (x - a)$, then x is said to be incongruent $a\ (mod\ n)$ and written as $x \not\equiv a\ (mod\ n)$.

Let $x \equiv a \ (mod\, n)$. Then a is said to be a residue of $x \ mod\, n$. Again if $0 = \leq a \leq n-1$ then a is said to be the least residue of $x \ mod\, n$.

For a given residue a the set of all integers x congruent to $a (mod\, n)$ is called the class of residues $mod\, n$. The difference of any two elements of this class is divisible by n. Let, $n = 4, a = 3$. Then if $x \equiv 3 \ (mod\, 4), 4|(x-3)$, $i.e.\, x - 3 = 4q, x = 4q + 3$. By taking different integer values of q we obtain the set of values of x as $\{..., -17, -13, -9, -5, -1, 3, 7, 11, 15, 19, ...\}$

We denote this set by $\bar{3}$. Similarly,

$$\bar{0} = \{..., -4, 0, 4, 8, ...\}$$

$$\bar{1} = \{..., -7, -3, 1, 5, 9, ...\}$$

$$\bar{2} = \{..., -6, -2, 2, 6, 10, ...\}$$

Similarly, if $m = 5$ we get

$$\bar{0} = \{..., -15, 10, -5, \ 0, 5, 10, ...\}$$

$$\bar{1} = \{..., -14, -9, -4, 1, 6, 11, ...\}$$

$$\bar{2} = \{..., -13, -8, -3, 2, 7, 12, ...\}$$

$$\bar{3} = \{..., -12, -7, -2, 3, 8, 13, ...\}$$

$$\bar{4} = \{..., -11, -1, -3, 4, 9, 14, ...\}$$

It is clear that for a given n, the n number of residue classes are as follows:

$$\bar{0} = \{..., -3n, -2n, -n, 0, n, 2n, 3n, ...\}$$

$$\bar{1} = \{..., -3n + 1, -2n + 1, -n + 1, 1, n + 1, 2n + 1, 3n + 1, ...\}$$

$$\bar{2} = \{..., -2n + 2, -2n + 2, -n + 2, 2, n + 2, 2n + 2, 3n + 2, ...\}$$

.......…....

$$\bar{r} = \{..., -3n + r, -2n + r, -n + r, r, n + r, 2n + r, 3n + r, ...\}$$

$$\overline{n-1} = \{..., -2n - 1, -n - 1, -1, n - 1, 2n - 1, 3n - 1, ...\}$$

Definition

Let H be a subgroup of a group G and let $a \in G$. Then the set

$$Ha = \{ha: h \in H\}$$

is called a right coset of H in G. Left cosets are defined similarly. Note that if G is an abelian group then the left cosets and right cosets are identical.

Theorem 3.14

For all $a \in G$

$Ha = \{x: x \in G \text{ and } a \equiv x \bmod H\}$.

Proof: For $a \in G, Ha = \{ha: h \in H\}$

Let $[a] = \{x: a \equiv x \bmod H\}$. We shall first show that $Ha \subset [a]$. For $h \in H, a(ha)^{-1} = a(a^{-1}h^{-1}) = h^{-1}$. So

$$a(ha)^{-1} \in H,$$

$$\text{i.e. } a \equiv (ha) \bmod H.$$

So $ha \in [a]$. Hence for every $h \in H$ we have $Ha \subset [a]$.

Now suppose that $x \in [a]$.

Then $ax^{-1} \in H$ and so $(ax^{-1})^{-1} = xa^{-1} \in H, i.e. xa^{-1} = h$ for some $h \in H$. Multiplying both sides by a on the right, we get $x = ha$ so $x \in Ha$. Hence $[a] \subset Ha$. Thus we have $[a] = Ha$.

Remark

These equivalence classes yield a decomposition of G into disjoint classes. Any two right cosets of H in G are either identical or have no element in common. We have the following theorem.

Theorem 3.15

There is a 1–1 correspondence between any two right cosets of H in G.

Proof: Let Ha and Hb be any two right cosets of H in G.

Define mapping $f: Ha \to Hb$ by

$$f(ha) = hb, h \in H$$

Now f is 1–1, for if $f(h_1a) = f(h_2a)$ then $h_1b = h_2b$ and so $h_1 = h_2$. f is clearly onto, for $hb \in Hb \Rightarrow ha \in Ha \text{ and } f(ha) = hb$.

Remark

If H is a subgroup of a finite group G then it follows from the above theorem that any two right cosets of H in G have the same number of elements. $H = He$ is itself a right coset of H in G. $o(H)$ denotes the order of H. It now follows that $o(H) = o(He)$ and hence the number of elements in a right coset of H in G is equal to $o(H)$. We have the following theorem.

Theorem 3.16 (Lagrange's Theorem)

If G is finite group and H is a subgroup of G then $o(H)$ is a divisor of order of G.

Proof: Let G be a finite group and H a subgroup. Let k be the number of distinct right cosets of H in G. Any two distinct right cosets have no element in common and the number of elements in each is equal to $o(H)$.

Therefore $k\ o(H) = o(G)$ and this completes the proof.

Definition

Let H be a subgroup of G. Index of H in G is the number of distinct right cosets of H in G and is denoted by $i_G(H)$. If G is a finite group then by Lagrange's theorem we have,

$$i_G(H) = \frac{o(G)}{o(H)}$$

Corollary 1

If G is a finite group and $a \in G$, then $o(a)|o(G)$.

Proof: Consider the subgroup $\{e, a, a_2, a_3, \dots\} = (a)$

We claim that the number of $(a) = o(a)$. Since $a^{o(a)} = e$ the subgroup has atmost $o(a)$ elements.

If the number of elements is less than $o(a)$, then $a^i = a^j$ for some integers $0 \leq i < j < o(a)$.

So $a^{j-i} = e$ when $0 < j - i < o(a)$ and this is a contradiction.

Thus the cyclic subgroup generated by a has $o(a)$ elements. By Lagrange's theorem $o(a)|o(G)$.

Corollary 2

If G is a finite group and $a \in G$ then $a^{o(G)} = e$.

Proof: We have $o(a)|o(G)$. So $o(G) = mo(a)$ for some $m \in N$. Therefore $a^{o(G)} = (a^{o(a)})^m = e^m = e$.

Corollary 3

If G is a finite group whose order is a prime number p then G is a cyclic group.

Proof: We claim that G has no nontrivial subgroup H. For, $o(H)$ must divide $o(G) = p$ so that,

$$o(H) = p \text{ or } o(H) = 1$$

$$\text{If } o(H) = 1 \text{ then } H = \{e\}$$

If $o(H) = p$ then $H = G$

Suppose that $a \neq e \in H$ and $H = (a)$. H is a subgroup of G, $H \neq \{e\}$ since $a \neq e \in H$.

Thus $H = G$. From this it follows that G is cyclic and that every element of G is a power of a.

Example 3.7

(a) If H is a subgroup of all even integers of the group Z of all integers under addition, then cosets of H are as follows:

$$H = H + 0 = \{\dots, -6, -4, -2, 0, 2, 4, 6, \dots\}$$
$$H + 1 = \{\dots, -5, -3, -1, 1, 3, 5, 7, \dots\}$$
$$H + 2 = \{\dots, -4, -2, 0, 2, 4, 6, \dots\}$$
$$H + 3 = \{\dots, -3, -1, 1, 3, 5, \dots\}$$
$$H + 4 = \{\dots, -2, 0, 2, 4, \dots\}$$
$$H + 5 = \{\dots, -1, 1, 3, 5, 7, \dots\}$$

It should be noted that,

$H = H + 2 = H + 4 = \cdots$

$H + 1 = H + 3 = H + 5 = \cdots$

Therefore two distinct cosets are H *and* $H + 1$ and $Z = H \cup (H + 1)$

It should also be noted here that since Z is an abelian group under addition the left cosets and the right cosets of H are identical.

(b) Find out the subgroups and their order of the group of all permutations of the three element set $\{a, b, c\}$.

This set has $3! = 6$ permutations which are as follows:

$$I = P_0 = \begin{pmatrix} a & b & c \\ a & b & c \end{pmatrix}, P_1 = \begin{pmatrix} a & b & c \\ a & c & b \end{pmatrix}, P_2 = \begin{pmatrix} a & b & c \\ b & a & c \end{pmatrix}$$

$$P_3 = \begin{pmatrix} a & b & c \\ b & c & a \end{pmatrix}, P_4 = \begin{pmatrix} a & b & c \\ c & a & b \end{pmatrix}, P_5 = \begin{pmatrix} a & b & c \\ c & b & a \end{pmatrix}$$

Now $\{P_0, P_1, P_2, P_3, P_4, P_5\}$ is a group under multiplication of permute-tions and its order is 6.

The sets $\{P_0, P_1\}$, $\{P_0, P_2\}$ and $\{P_0, P_5\}$ are subgroups of the above group each of order 2. The set $\{P_0, P_3, P_4\}$ is a subgroup of order 3. The group itself is a subgroup of order 6 and $\{P_0\}$ is a subgroup of order 1. The group does not have any subgroup of order 4 or 5 (see Lagrange's theorem). All the subgroups are of order 1, 2, 3, and 6, each being a divisor of 6 (the order of the group).

Definition

Let G be a group and H a subgroup of G. H is said to be a normal subgroup of G if every left coset of H in G is a right coset of H in G. If H is normal subgroup of G it is denoted by $H \Delta G$. In other words, if for each $a \in G$ we have $Ha = aH$ then $H \Delta G$.

Remark

(a) Note that $Ha = aH$ does not mean that $ha = ah$ for every $h \in H$. It really means that for every $h \in H$ there exists some $h_1 \in H$ such that $ha = ah_1$.
(b) The trivial subgroups $\{e\}$ and G are always normal subgroups.
(c) A group may not have any nontrivial normal subgroup. Such groups are called simple groups.
We now give an equivalent characterization of a normal subgroup.
If H is a subgroup of a group G and if $a \in G$ then we write,

$$aHa^{-1} = \{aha^{-1} : h \in H\}$$

Theorem 3.17

(i) H is a normal subgroup of G if and only if $aHa^{-1} \subset H$ for each $a \in G$.
(ii) Moreover $H \Delta G$ if and only if $aHa^{-1} = H$ for each $a \in G$.

Proof: Assume that $H \Delta G$. Therefore for each $a \in G$ we have $aH = Ha$. Let $h_1 \in H$, i.e., $ah_1a^{-1} \in aHa^{-1}$. Since $aH = Ha$ we can choose $h_2 \in H$ such that $ah_1 = h_2a$, so that $ah_1a^{-1} = (h_2a)a^{-1} = h_2 \in H$. Therefore $aHa^{-1} \subset H$. Further,

$$a^{-1}Ha = a^{-1}H(a^{-1})^{-1} \subset H$$

Now since $a^{-1}Ha \subset H, H = a(a^{-1}Ha)a^{-1} \subset aHa^{-1} \subset H$ and hence $aHa^{-1} = H$.

Conversely assume that for each $a \in G$, $aHa^{-1} \subset H$. We have to show that $H \Delta G$, in other words $aH = Ha$. Now for any $h \in H, ah \in aH$. Since $aHa^{-1} \subset H, aha^{-1} \in H$. So we can choose some $h_1 \in H$ such that,

$$aha^{-1} = h_1, i.e\ ah = h_1a.$$ Since $h_1a \in Ha, ah \in Ha$.

Since $a^{-1} \in G$ we have,

$$a^{-1}Ha = a^{-1}H(a^{-1})^{-1} \subset H$$

So we can choose h and h_1 in such a manner that,

$$a^{-1}h(a^{-1})^{-1} = h_1 \text{ or } a^{-1}ha = h_1 \text{ or } ha = ah_1$$

Since $ha \in Ha$ so $ha \in aH$. Thus $Ha \subset aH$ and hence $Ha = aH$. Now if $aHa^{-1} = H$ for every $a \in G$ certainly $aHa^{-1} \subset H$ and therefore the result follows.

Example 3.8

(a) If G is a group then cent G, the centre of G,is a normal subgroup of G. We have already demonstrated that Cent G is a subgroups of G. Note that if $c \in$ Cent G then from the definition of Cent G, it follows that $ca = ac$. Therefore, $aca^{-1} = (ca)a^{-1} = c(aa^{-1}) = c \in cent\ G$. Hence cent $G\Delta G$.

(b) If H_1 and H_2 are normal subgroups of G then $H_1 \cap H_2$ is a normal subgroup of G. We have proved earlier that if $H_1$and H_2 are subgroups of G then $H_1 \cap H_2$ is also a subgroup of G. Now let $h \in H_1 \cap H_2$. Then $h \in H_1$ and $h \in H_2$. Since H_1 and H_2 are normal subgroups, for each $a \in G, aha^{-1} \in H_1$ and $aha^{-1} \in H_2$. Therefore $aha^{-1} \in H_1 \cap H_2$ and thus $H_1 \cap H_2 \Delta G$.

(c) Every subgroup of an abelian group is normal. To show this, let G be an abelian group and let $H < G$. We have to show that if $h \in H$ then for every $a \in G, aha^{-1} \in H$. Let $h \in H$. Then for any $a \in G$,

$$aha^{-1} = (ah)a^{-1} = (ha)a^{-1} \text{ (since G is abelian)}$$
$$= h(aa^{-1}) = he = h \in H$$

However, the converse need not be true, in other words, if every subgroup of a group G is normal then G need not be abelian (Give an example).

Theorem 3.18

A subgroup $H$of G is a normal subgroup of G if and only if the product of two right cosets of H in G is again a right coset of H in G.

Proof: Let $H\Delta G$ and let $a, b \in G$. First of all note that for any subgroup H of G, $HH = H$. Now consider the product $(Ha)\,(Hb)$ of two right cosets Ha and Hb of H $in\ G$. Since $H \Delta G$, $aH = Ha$ and so,

$$(Ha)\,(Hb) = H(aH)b = H(Ha)b = HHab = Hab$$

Thus the product of two right cosets is again a right coset.

Conversely, let $H < G$ and let the product of any two right cosets of $H\ in\ G$ be a right coset of $H\ in\ G$.

For $a \in G$, Ha and Ha^{-1} are two right cosets of $H\ in\ G$ and hence $HaHH^{-1}$ is also a right coset of $H\ in\ G$. Now $eaea^{-1} \in HaHa^{-1}$, but $eaea^{-1} = aa^{-1} = e \in H$ and H is itself a right coset. Further if two right cosets have one element common, they must be identical, Hence for all $a \in G, H = HaHa^{-1}$. Let $h \in H$ and $a \in G$. Then,

$$\begin{aligned} aha^{-1} &= (ea)(ha^{-1}) \in (Ha)(Ha^{-1}) \\ &= H \end{aligned}$$

and this completes the proof.

We have observed earlier that if H and K are subgroups of a group G then HK need not be a subgroup, it is, however, a subgroup if and only if $HK = KH$ or in particular if G is abelian. We shall now show that if one of them is a normal subgroup, then the product is a subgroup and if both are normal, the product is a normal subgroup.

Theorem 3.19

Let G be a group, $N \Delta G$ and $H < G$. Then $NH < G$. If further $H \Delta G$, then $NH \Delta G$.

Proof: Let $nh \in NH$ and $n_1h_1 \in NH$.

$$\begin{aligned} \text{Then } (nh)(n_1h_1)^{-1} &= (nh)(h_1{}^{-1}n_1{}^{-1}) \\ &= n(hh_1{}^{-1})n_1{}^{-1} \\ &= nh_2n_1{}^{-1} \text{ (say } hh_1{}^{-1} = h_2) \end{aligned}$$

Since $H < G$, $hh_1{}^{-1} = h_2 \in H$. Also, $h_1 \in H$. Since $N\Delta G, Na = aN$ for each $a \in G$. Now $h_2 \in H \subset G$, so $Nh_2 = h_2N$. Again if $n_1 \in N, n_1{}^{-1} \in N$. We can now choose $n_2 \in N$ such that $n_2h_2 = h_2n_1{}^{-1}$.

Therefore,

$$\begin{aligned} (nh)(n_1h_1)^{-1} &= n(h_2n_1)^{-1} = n(n_2h_2) \\ &= (nn_2)h_2 - n_3h_2 \in NH \end{aligned}$$

Since $nn_2 = n_3 \in N$. Therefore it follows that $NH \subset G$.

We shall now show that NH is normal in G. Let $nh \in NH$. Then for any $a \in G$,

$$\begin{aligned} a(nh)a^{-1} &= (an)\,(ha^{-1}) \\ &= \big((an)e\big)(ha^{-1}) \\ &= (an)(a^{-1}a)(ha^{-1}) \\ &= (ana^{-1})(aha^{-1}) \in NH \end{aligned}$$

Since N and H are normal subgroups, $ana^{-1} \in N$ and $aha^{-1} \in H$. Thus $NH \,\Delta\, G$ and this completes the proof.

We have already proved that the intersection of any two subgroups of a group G is again a subgroup. We shall now show that if one of them is a normal subgroup, the intersection is a normal subgroup.

Theorem 3.20

Let G be a group, $H < G$ and $N \Delta G$. Then $H \cap N \Delta H$.

Proof: Since $H \cap N \subset H$ and $H \cap N < G$. So that $H \cap N < H$. Let $a \in H$ and $h \in H \cap N$. Then $h \in H$ and $h \in N$. Since $N\Delta G$, $aha^{-1} \in N$. Since $h \in H, a \in H$ we have $ah \in H$.

Further $a^{-1} \in H$, so $(ah)a^{-1} \in H$. Therefore $(aha^{-1}) \in H \cap N$ and this shows that $H \cap N$ is a normal subgroup of H.

We have already observed that for a normal subgroup of a group G the product of two right cosets is again a right coset. We shall now demonstrate that we can use this product to make the collection of right cosets into a group, which will be called the quotient group or the factor group. This type of structures which are called quotient structures are very important and useful in mathematics.

If H is a normal subgroup in a group G, G/H denotes the collection of all right cosets of H in G, i.e.

$$G/H = \{Ha : a \in G\}$$

Recall that for $X = Ha \in G/H$ and $Y = Hb \in G/H$ for some $a, b \in G$ we define $XY = NaNb = Nab \in G/H$.

We have.

Theorem 3.21

If G is a group and $N\Delta G$, then G/H is a group (called the quotient group or factor group) under the product defined as above.

Proof: Let $X = Na, Y = Nb$ and $Z = Nc$ be in G/H with $a, b, c \in G$.

Then

$$(XY)Z = (NaNb)Nc = NabNc = N(ab)c = Na(bc)$$
$$= NaNbc = Na(NbNc) = X(YZ)$$

and hence associativity holds. Now $H = He \in G/H$ is the identity element of G/H, for if $X = Ha \in G/H$ for some $a \in G$, then

$$XH = HaHe = Hae = Ha = X$$

and similarly $HX = X$.

Suppose $X = Ha \in G/H$ where $a \in G$.

Then $Ha^{-1} \in G/H$ and is the inverse of Ha, since $HaHa^{-1} = Haa^{-1} = He = H$, and similarly $Ha^{-1}Ha = H$.

This completes the proof of the theorem.

Remark

(a) If G is a finite group and H is a normal subgroup of G, then $o(G/H) =$ the number of elements in $G/H =$ the number of distinct right cosets of H in $G = i_G(H) =$ the index of H in G. Therefore by using Lagrange's theorem we have,

$$o(G/H) = \frac{o(G)}{o(H)}$$

(b) If G is abelian and $H < G$, then the quotient group $\frac{G}{H}$ is also abelian. To prove this let G be abelian and let $H < G$. Then H is also abelian and $H \Delta G$. Let $X = Ha, Y = Hb$ be in $\frac{G}{H}$ for $a, b \in G$. Then $XY = HaHb = Hab = Hba$ (since G is abelian) $= HaHb = YX$ and this proves the result.

3.3 HOMOMORPHISM

In this section we shall discuss about the mappings from one group into another which preserve the algebraic operations. Such mappings are called homomorphisms and if there is one central idea which is common to all aspects of algebra, it is the concept of homomorphism. The ideas and results here are closely connected with and use the concept of normal subgroups.

Definition

Let G and $\bar{G}$ be groups and let $f: G \rightarrow \bar{G}$. f is said to be a homomorphism from G into $\bar{G}$ if for all $a, b \in G, f(ab) = f(a)f(b)$.

A homomorphism f from G into $\bar{G}$ is called an isomorphism if f is 1–1 and two groups G and $\bar{G}$ are said to be isomorphic if there exists an isomorphism from G onto $\bar{G}$. A homomorphism of a group G into itself is called an endomorphism. An isomorphism of a group G into itself is called an automorphism. If f is a homomorphism from G into $\bar{G}$ then the kernel of f, denoted by K_f, is defined by

$$Kf = \{x \in G: f(x) = \bar{e} = \text{the identity element of } \bar{G}\}.$$

Remark

Note that in the definition of homomorphism for $a, b \in G$, by ab we mean the composition in G and by $f(a)f(b)$ we mean the composition in $\bar{G}$. For convenience we use the same symbol and it should not create any confusion.

Example 3.9

(a) Note that the mapping $f: G \rightarrow G$ defined by $f(x) = \bar{e}$ for all $x \in G$ is trivially a homomorphism (in fact an endomorphism) with $K_f = G$.

(b) The map $f: G \rightarrow G$ defined by $f(x) = x$ for all $x \in G$ is a homomorphism (an automorphism) with $K_f = \{e\}$.

(c) Let G be the group C of all complex numbers under addition. Then the map $f: C \rightarrow C$ define by $f(z) = \bar{z}$ for all $z \in C$ is an automorphism.

(d) Let $G = Z$, the group of integers under addition. Then $f: G \rightarrow G$ defined by $f(x) = 2x$ is a homomorphism.

(e) Let G= the group of real numbers under addition and $\bar{G}$ = the group of nonzero real numbers under multiplication. Then $f: G \rightarrow \bar{G}$ defined by $f(a) = 2^a$ is a homomorphism.

(f) Let G be a group and $H < G$. Then the inclusion map $i: H \rightarrow G$ defined by $i(a) = a$ for all $a \in H$ is a homomorphism from H into G.

(g) Let G = the group of nonzero reals under multiplication and $\bar{G} = \{1, -1\}$. Then $f: G \rightarrow \bar{G}$ defined by,

$$f(x) = \begin{cases} 1, x \text{ positive} \\ -1, x \text{ negative} \end{cases}$$

is a homomorphism.

(h) Let G = the group of integers under addition and $\bar{G}$ = the group of integers under addition modulo n. Define f by $f(x)$ = remainder of x on division by n. Then f is a homomorphism.

(i) G = the group of positive real numbers under multiplication and $\bar{G}$ = the group of all real numbers under addition. Define $f : G \to \bar{G}$ by $f(x) = \log_{10} x$. Then f is a homomorphism and it is also 1-1 and onto.

(j) Let G be the group of all 2×2 matrices $\begin{pmatrix} a & b \\ c & d \end{pmatrix}$ such that $ad - bc \neq 0$, under matrix multiplication. Let $\bar{G}$ be the group of all nonzero real numbers under multiplication. Define $f : G \to \bar{G}$ by

$$f\left(\begin{pmatrix} a & b \\ c & d \end{pmatrix}\right) = ad - bc$$

then f is a homomorphism from G onto $\bar{G}$.

(k) Let G= the group of all integers under addition and $\bar{G}$ = the group of positive real numbers under multiplication. Then $f: G \to \bar{G}$ defined by $f(x) = e^x$ is an isomorphism.

(l) Any finite cyclic group G of order n is isomorphic to the group of residue class mod n under addition mod n.

If G is a finite cyclic group of order n with generator a and $\{\bar{0}, \bar{1}, \ldots, \overline{n-1}\}$ is the group of residue class mod n, then the map given by $f(a^r) = \bar{r}$ for $r = 0,1,\ldots,n-1$, defines an isomorphism. For example, $G = \{1, -1\}$ under multiplication is isomorphic to $\bar{G} = \{0, 1, 2, 3\ (mod) 4\}$ under addition.

(m) Any infinite cyclic group G is isomorphic to Z, the group of integers under addition. If a is a generator of G then the map f given by,

$f(a^k) = k$

is an isomorphism.

(n) The group Z of integers under addition is not isomorphic to the group of nonzero rational numbers under multiplication. (Prove this!)

Theorem 3.22

Let f be a homomorphism from G into $\bar{G}$. Then

(i) $f(e) = \bar{e}$ where $\bar{e}$ is the unit element of $\bar{G}$,

(ii) $f(a^{-1}) = f(a)^{-1}$ for each $a \in G$

Proof:

(i) For each $a \in G, f(a)\bar{e} = f(a) = f(ae) = f(a)f(e)$ and hence $f(e) = \bar{e}$.

(ii) For each $a \in G,\ f(a)f(a^{-1}) = f(aa^{-1}) = f(e) = e = f(a)f(a)^{-1}$ and hence $f(a^{-1}) = f(a)^{-1}$.

We shall now show that the kernel of a homomorphism has very interesting property, namely it is a normal subgroup.

Theorem 3.23

Let f be a homomorphism from G into $\bar{G}$ with the kernel K_f, Then K_f is a normal subgroup of G.

Proof: First of all note that $K_f \neq \varphi$ since $e \in K_f$, we now check that $K_f \in G$. For this purpose, let $a, b \in K_f$, Then $f(a) = \bar{e}$ and $f(b) = \bar{e}$. Now $f(ab^{-1}) = f(a)\,f(b^{-1}) = \bar{e}f(b)^{-1} = \overline{ee}^{-1} = \overline{ee} = e$. Hence, $ab^{-1} \in K_f$, and thus $K_f < G$. For the normality of K_f, we have to show that for every $a \in G$ and $k \in K_f$, $aka^{-1} \in K_f$. Now let $a \in G$ and $k \in K_f$. Then,

$$f(aka^{-1}) = f(a)f(k)f(a^{-1}) = f(a)\bar{e}f(a)^{-1} = f(a)f(a)^{-1} = \bar{e}$$

and this completes the proof.

We now consider the special case of homomorphism namely isomorphism and show that.

Theorem 3.24

The relation of isomorphism in the set of all groups is an equivalence relation.

Proof: If G and $\bar{G}$ are isomorphic we write this as $G \approx \bar{G}$. We have to show that:

(i) $G \approx G$ (reflexivity)

(ii) $G \approx \bar{G}$ implies $\bar{G} \approx G$ (symmetry)

(iii) $G \approx \bar{G}, \bar{G} \approx \bar{\bar{G}}$ implies $G \approx \bar{\bar{G}}$(transitivity).

For each $x \in G$ define $f: G - G\ by\, f(x) = x$. Then f is an isomorphism of G onto itself and hence $G \approx G$.

Let $G \approx \bar{G}$. Then there exists an isomorphism f from G onto $\bar{G}$. Since f is 1–1 and onto, the map $f^{-1}: \bar{G} \to G$ exists and is 1–1 and onto. Let $a, b \in \bar{G}$ and let $f(x) = a, f(y) = b$ for $x, y \in G$. Now $x = f^{-1}(a), y = f^{-1}(b), f(xy) = f(x)f(y) = ab$ and so $f^{-1}(ab) = xy = f^{-1}(a)f^{-1}(b)$. Then f^{-1} is also a homomorphism and so $\bar{G} \approx G$.

For the transitivity, let $G \approx \overline{G}$ and $\bar{G} \approx \bar{\bar{G}}$. Thus there exist isomorphisms f and g from G onto $\overline{G}$ and from $\bar{G}$ onto $\bar{\bar{G}}$ respectively. Since f and g are one-one and onto, the composition g of : $G \to \bar{\bar{G}}$ is 1–1 and onto. Further from $x, y \in G$,

$$(g \text{ of})(xy) = g(f(xy)) = g(f(x)f(y)) = g(f(x))g(f(y))$$

Thus $g \, of$ is a homomorphism and hence $G \approx \bar{\bar{G}}$. This completes the proof.

Thus it follows from the above theorem that when two groups are isomorphic, then they are indistinguishable in some sense and we can treat them equal. They differ only in that their elements are labeled differently. If we know a given computation in one group, we can carry out the analogous computation in another group to which it is isomorphic.

Theorem 3.25 (Fundamental theorem of Homomorphism)

Let φ be a homomorphism of G into $\bar{G}$, with kernel K. Then $G/K \approx \bar{G}$.

Proof:

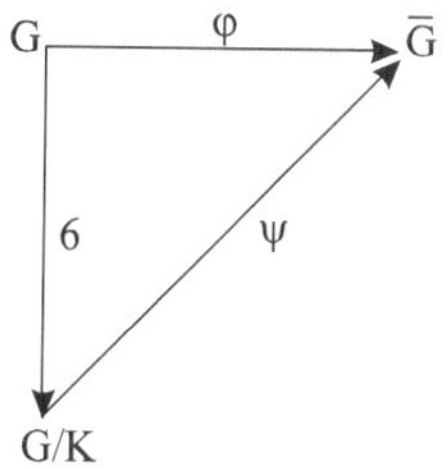

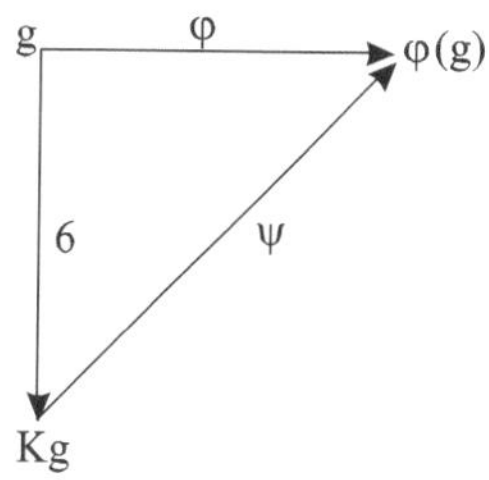

Define the mapping $\psi: G/K \to \bar{G}$ as follows. If $X \in G/K, X = Kg$ then $\psi(X) = \phi(g)$. First of all note that ψ is well defined,

$$\text{If } X = Kg = Kg', g, g' \in G$$

$$\psi(X) = \phi(g) \text{ and } \psi(X) = \phi(g')$$

In order that ψ makes sense it must be necessary that $\phi(g) = \phi(g')$. Suppose $Kg = Kg'$. Then $g = kg', k \in K$.

Therefore $\phi(g) = \phi(kg') = \phi(k)\phi(g') = \bar{e}\phi(g') = \phi(g')$

is onto: For, if $\bar{x} \in \bar{G}, \bar{x} = \phi(g), g \in G$ since ϕ, is onto, we have

$$\bar{x} = \phi(g) = \psi(X) = \psi(Kg)$$

Thus ψ is onto.

ψ is a homomorphism of $\frac{G}{K}$ onto $\bar{G}$. If $X, Y \in \frac{G}{K}, X = Kg, Y = Kf, g, f \in G$ then $XY = KgKf = Kgf$.

Hence

$\psi(XY) = \psi(K_{gf}) = \phi(gf) = \phi(g)\phi(f)$, since ϕ is a homomorphism of G onto $\bar{G}$. But $\psi(X) = \psi(Kg) = \phi(g)$ and $\psi(Y) = \psi(K_f) = \phi(f)$

Therefore, $\psi(XY) = \psi(X)\psi(Y)$

Finally we show that kernel of ψ = unit element in G/K, The unit element of $G/K = K_e = K$.

Let $Ker\,\psi = Kg$.

Now $\bar{e} = \psi(K_g) = \phi(g)$

So $g \in Ker\phi = K$. Hence $K_g = K$, and this completes the proof.

Remark

If N is a normal subgroup of G then it follows from above theorem that upto isomorphism G/N is a homomorphic image of G. Therefore there exists a 1–1 correspondence between homomorphic images of G and normal subgroups of G.

Let G be a group and let $A(G)$ be the set of all automorphisms of G onto itself.

Note that $A(G)$ is nonempty, for I the identity map, belongs to $A(G)$, we have.

Theorem 3.26

$A(G)$ is a group, the composition being the usual composition of maps.

Proof: First of all note that $A(G)$ is closed under the composition. Let $T_1, T_2 \in A(G)$.

We shall show that $T_1T_2 \in A(G)$.

Let $x, y \in G$. Since $T_1, T_2 \in A(G), T_1(xy) = (T_1x)(T_1y)$ and $T_2(xy) = (T_2x)(T_2y)$.

$$\text{Now } T_1T_2(xy) = T_1(T_2(xy)) = T_1((T_2x)(T_2y)) = (T_1(T_2x))(T_1(T_2y))$$
$$= (T_1T_2(x))\,(T_1T_2(y))$$

We have only to show that if $T \in A(G)$ then $T^{-1} \in A(G)$.

Now if

$$T((T^{-1}x)T^{-1}y) = T(T^{-1}x)(T(T^{-1}\,y)) = ((TT^{-1})(x))\,(TT^{-1}(y)$$
$$= I(x)(Iy) = xy$$

This means that:

$$T^{-1}(xy) = (T^{-1}x)(T^{-1}y)$$

and completes the proof.

3.3.1 Discussions

We have just demonstrated that $A(G)$ the set of all automorphisms of a group G into itself is nonempty and is a group. It is now a pertinent question whether $A(G)$ is a nontrivial group or whether it consists of I only; in other words do there exist nontrivial automorphisms. If G is a group consisting of two elements only, then it is left to the reader as an exercise to verify that $A(G) = \{I\}$. But we shall now demonstrate that if G has more than two elements, then $A(G)$ is nontrivial.

Firstly, let G be abelian and let there exist an element $x_0 \in G$ such that $x_0 \neq x_0^{-1}$.

Let $T: G \to G$ be defined by $T(x) = x^{-1}$ for all $x \in G$. Then T is an automorphism and,

$$T \neq I.$$

Now let G be any group, not necessarily abelian. For given $g \in G$ define a map by $Tg: G \to G$ by,

$$T_g(x) = g^{-1}xg \text{ for all } x \in G$$

Observe that T_g is an automorphism of G onto G. For given $y \in G$, let $x = gyg^{-1}$.

Then

$$T_g(x) = g^{-1}xg = g^{-1}(gyg^{-1})g$$
$$= (g^{-1}g)y(g^{-1}g) = y$$

and hence T is onto.

If $T_g(x) = T_g(y)$ then $g^{-1}xg = g^{-1}yg$ and hence $x = y$. Thus Tg is 1–1.

Further for $x, y \in G$.

$$T_g(xy) = g^{-1}(xy)g = g^{-1}(xgg^{-1}y)g$$
$$= (g^{-1}xg)(g^{-1}yg) = T_g(x)T_g(y)$$

Thus T_g is a homomorphism and hence it is an automorphism of G onto itself. If G is nonabelian, then there exists a pair of elements $a, b \in G$ such that

$ab \neq ba$, then $T(a)b = a^{-1}ba \neq b$ and so $Ta \neq I$. We now make the following definition.

Definition

For an element g in a group, G the automorphism Tg from G onto G defined by $T_g(x) = g^{-1}xg$ is called the inner automorphism corresponding to g. We write $I(G)$ for the set of all inner automorphisms of G, i.e.

$$I(G) = \{T_g \in A(G): g \in G\}$$

We have

Theorem 3.27

$I(G)$ is a group and $I(G) \approx G/Z$ where Z is the centre of G.

Proof: It is left to the reader to verify that $I(G)$ is a subgroup of $A(G)$ and hence $I(G)$ is a group.

To show that $I(G) = G/Z$ define a map $\psi: G \rightarrow A(G)$ by,

$$\psi(g) = T_g \text{ for every } g \in G$$

Then it can be easily checked that ψ is a homomorphism from G into $A(G)$, whose image is $I(G)$. Now we claim that the Kernel of ψ is Z. Suppose it is K and let $g_0 \in K$. Then $\psi(g_0) = I$ i.e. $Tg_0 = I$. This means that for any $x \in G, Tg_0(x) = x$. However, $Tg_0(x) = g_0{}^{-1}xg_0$ and so $x = g_0{}^{-1}xg_0$ for all $x \in G$, i.e. $g_0x = xg_0$ for all $x \in G$. Thus, $K \subset Z$. Now if $z \in Z$, then $T_zx = z^{-1}xz = z^{-1}(zx) = x$, so $T_z = I$ and hence $z \in K$, showing that $Z \subset K$. Therefore, $K = Z$.

Observe that ψ is a homomorphism of G into $A(G)$ with image $I(G)$ and Kernel Z.

Now applying the Fundamental theorem of homomorphism we get the required conclusion.

Theorem 3.28 (Cayley)

Every group is isomorphic to a subgroup of $A(S)$ for some appropriate S where $A(S)$ denotes the group of all 1–1 and onto mappings from a set S onto itself.

Proof: Let G be a group. Take $S = G$. For $g \in G$ define $T_g: S = G \rightarrow S = G$ by,

$$T_g(x) = xg \text{ for every } x \in G$$

If $y \in G$ then $y = (yg^{-1})g = T_g(yg^{-1})$, so that T_g is onto. If $x, y \in S$ and $T_g(x) = T_g(y)$, then $xg = yg$ and hence $x = y$. Therefore, T_g, is 1–1, and hence for every $g \in G, T_g \in A(S)$. Now define a map $\psi: G \to A(S)$ by,

$$\psi(g) = T_g$$

for all $g \in G$

Observe that ψ is a homomorphism. Let K be the kernel of ψ and let $g_0 \in K$. Then,

$\psi(g_0) = Tg_0$ is the identity map on S.

So for all $x \in G$, and in particular, for $e \in G$,

$$e = eTg_0 = eg_0 = g_0$$

Thus $g_0 = e$ and hence $K = \{e\}$.

Now it follows that ψ is an isomorphism of G into $A(S)$ and this proves the theorem.

3.4 SYLOW'S THEOREM

Lagrange's theorem tells that in a finite group the order of the subgroup divides the order of the group. The converse is not necessarily true. There are very few theorems which show the existence of subgroups of prescribed orders in any finite group. The most fundamental and very much useful is a famous theorem due to Sylow. We first prove.

Theorem 3.29 (Cauchy theorem for abelian groups)

If the order of a finite abelian group is divisible by a prime p then there exists an element of order p in that group.

Proof: We shall prove this theorem by the method of induction.

Assume that the theorem holds for each abelian group whose order is less than $O(G)$.

Trivially, the theorem holds for the group having single element. It is true as well for the groups having two and three elements as they are primes (if the order of G is a prime then it must be cyclic and its order is nothing but p. Hence all elements other than the identity have the order p).

Let H be a subgroup of G such that $G \neq H \neq \{e\}$. There are two cases: either $p|o(H)$ or $p \dagger o(H)$. If $p \mid o(H) < o(G)$ then by the induction hypothesis we have $b \in H$ such that $b^p = e$. Now consider the case when $p \dagger o(H)$. As G is abelian H is a normal subgroup and G/H is an abelian group. $p \dagger o(H)$ and $p|o(G) \Rightarrow p|\frac{o(G)}{o(H)} = o(G/H) < 0(G)$.

So, by induction hypothesis, we have a coset H_b, such that $(H_b)^p = H^p{}_b = H \Rightarrow b^p \in H$ but $b \notin H$. So $(b^p)^{o(H)} = e$, i.e. $b^{o(H)p} = e$. Let $a = b^{o(H)} = e. (H^b)^{o(H)} = H = (H^b)^p$. But $p \dagger o(H) \Rightarrow H = H_b \Rightarrow b \in H$ which is a contradiction. So $a \neq e$.

Theorem 3.30 (Sylow's theorem for abelian groups)

If G is a finite abelian group such that for some prime $p, p^k|o(G)$ but $p^{k+1} \dagger o(G)$, then G has a unique subgroup of order p^k.

Proof: If $k = 0$ the trivial subgroup $\{e\}$ is the required subgroup. If $k = 1$ then we have an element of order p which generates our regenerated subgroup. For other cases $p|o(G) \Rightarrow \exists\, a \in G$ such that $a \neq e$ but a^p =e. Let $H = \{x \in G: x^{p^n} = e \text{ for some integer } n\}$. $a^p = e \Rightarrow a \in H$ and hence $H \neq \{e\}$. Again if $x, y \in H$ and $x^{p^n} = e, yx^{p^m} = e$ then $(xy)^{p^{n+m}} = x^{p^{n+m}}$. $y^{p^{n+m}} = e \Rightarrow xy \in H$. So H is a subgroup of G. As $a \in H$ we have $o(a)|\, o(H) \Rightarrow p\,|o(H)$. If $q \mid o(H)$ where q is some prime other than p then we have $h \in H$ such that $h^q = e$ but $h \neq e$ since $h \in H$ where $h^{p^n} = e$ for some n. Since p^n and q are relatively prime (q is prime) we have integers s and t such that $sp^n + tq = 1$. So $c = c^1 = c^{sp^n+tq} = c^{sp^n} = c^{tq} = e \Rightarrow c = e$ which is a contradiction.

So no prime other than p divides $o(H)$. Thus we can claim that $o(H) = p^m$ for some integer m such that $0 < m \leq k$. If $m < k$ then $p|o(G/H) \Rightarrow \exists Hx \in G/H$ such that $Hx \neq H$ and $(Hx)^p = H$. But $H = Hn = (Hx)^{p^n} \Rightarrow x^{p^n} \in H \Rightarrow e = (x^{p^n})^{o(H)} = (x^{p^n})^{p^m} = x^{p^{n+m}}$. So $x \in H$ which is a contradiction. So $m = K$ and H is the required subgroup of order p^k. To prove the uniqueness we assume that $\exists K$, a subgroup of G whose order is also p^k and $H \neq K$. As G is abelian we have $HK = KH \Rightarrow HK$ is a subgroup and

$$o(HK) = \frac{o(H)o(K)}{o(H \cap K)} = \frac{p^k p^h}{o(H \cap K)}$$

Since $H \neq K, o(H \cap K) < p^k$ but some power of p. So $o(HK) = p^l$ when $l > K$. Since HK is a subgroup of $G, o(HK) \mid o(G)$ but $p^{k+1} \dagger o(G)$. So $p^l \dagger p(G)$ for $l > K$ which is a contradiction. So H must be unique.

Theorem 3.31

If G is a finite group then the number of elements conjugate to an element a in G is the index of the normalizer of a in G.

From this we have,

$$o(G) = \sum_{a \in G} o(G)/o(N(a))$$

If we have,

$a \in Cent\ G$ then $N(a) = G$. So we can rewrite the above equation as,

$$o(G) = o(CentG) + \sum_{\substack{a \in c \\ a \notin CentG}} o(G)/o(N(a))$$

This is called the class equation of G.

Proof: Let N be the normalizer of an element a in G. The index of N in G is $o(G/N)$.

We will now define a mapping from the conjugate class of a to G/N and will show that the mapping is 1–1 and onto which will prove the theorem.

Let $x_1 x_2 \in G$ such that $x_1 a x_1^{-1} = x_2 a x_2^{-1}$

$\Rightarrow a = x_1^{-1} x_2 a x_2^{-1} x_1 = x_1^{-1} x_2 a (x_1^{-1} x_2)^{-1}$

$\Rightarrow a$ is conjugate to $x_1^{-1} x_2$

$\Rightarrow x_1^{-1} x_2 \in N$

$\Rightarrow x_1 \equiv x_2 (mod\ N)$

$\Rightarrow x_1$ and x_2 will be in same left coset of N. Conversely let us assume that x_1 and x_2 are in the same left coset bN of N in G. Then we have $x_1 = bn_1$ and $x_2 = bn_2$ for $n_1, n_2 \in N$. Now

$x_1 a\, x_1^{-1} = bn_1 a (bn_1)^{-1} = bn_1 a n_1^{-1} b^{-1} = bab^{-1}$ and

$x_2 a x_2^{-1} = bn_2 a (bn_2)^{-1} = bn_2 a n_2^{-1} b = bab^{-1}$

Set $x_1 a x_1^{-1} = x_2 a x_2^{-1}$ iff x_1, x_2 belong to the same left coset of N in G. Now we define a mapping f from $c(a)$ to G/N as follows:

$$f(xax^{-1}) \rightarrow xN$$

where $c(a)$ denotes the conjugate class of a.

From the 'iff' condition of the result already proved we can see that the mapping is 1–1 and onto. Hence the result.

Theorem 3.32 (Cauchy)

If a prime p divides the order of G then G has an element of order p.

Proof: We prove this theorem by induction. Only difference is when $p \nmid o(H)$ where H is any proper subgroup of group G, not equal to $\{e\}$. Hence, as G is not abelian in particular, H need not be normal. We proceed to complete the theorem that G has no non-trivial subgroup whose order is divisible by p. From the class equation we have,

$$o(G) = o(CentG) + \sum_{\substack{a \in G \\ a \notin CentG}} o(G)/o(N(a))$$

$$p | o(G) \text{ and } p \nmid o(N(a))$$

So $p | \sum_{\substack{a \in c \\ a \notin CentG}} o(G)/o(N(a))$

$$\Rightarrow p \left| \left[o(G) - \sum_{\substack{a \in c \\ a \notin CentG}} o(G)/o(N(a)) \right] \right.$$

$$\Rightarrow p | o(CentG)$$

Cent G is a subgroup of G and by our assumption it should be a proper subgroup of G. So Cent $G = G \Rightarrow G$ is abelian.

So by theorem 3.29 we have an element in G whose order is p.

Lemma 3.1

If p is a prime and if s, m are positive integers, then

$$p \nmid \binom{mp^s - 1}{p^s - 1}$$

Proof: We have $\binom{mp^s - 1}{p^s - 1} = \frac{(mp^s-1)(mp^s-2)...(mp^s-p^s+1)}{1.2.3...(p^s-1)}$

Let k be a positive integer such that $1 \leq k \leq p^s - 1$. Suppose $p^t | (mp^s - k)$ for some positive t. As $k < p^s$ we have $p^s | k$, so $p^s \nmid (mp^s - k)$ but $p^t | (mp^s - k) \Rightarrow t < s \Rightarrow p^t | mp^s \Rightarrow p^t \big| \big(mp^s - (mp^s - k)\big) \Rightarrow p^t \big| k$.

So in the expression for $\binom{mp^s - 1}{p^s - 1}$ the powers of p (if any) in the corresponding factor cancel out resulting in $p \nmid \binom{mp^s - 1}{p^s - 1}$.

Theorem 3.33 (Sylow's First Theorem)

If $p^t \mid o(G)$ for some prime p then G has a subgroup of order p^t.

Proof: Let $o(G) = mp^t$. Consider $\binom{mp^t}{p^t} = m\binom{mp^t - 1}{p^t - 1}$. So, if $p^r \mid m$ but $p^{r+1} \dagger m$ then we have $p^r \mid \binom{mp^t}{p^t}$ but $p^{r+1} \dagger \binom{mp^t}{p^t}$. Let P be the set of all subsets of G which have p^t elements. So $o(P) = \binom{mp^t}{p^t}$.

We define an equivalence relation of P as follows:

For $P_1, P_2 \in P, P_1 {\sim} P_2$ iff

$P_1 = P_2 g$ for some $g \in G$

The equivalence relation divides P into disjoint equivalence classes such that order of at least one equivalence class is not divisible by p^{r+1}. If not, then order of each equivalence class is divisible by p^{r+1}resulting, $o(P)$ is p^{r+1}, which is not true. So let $\{P_1, \ldots, P_k\}$ be an equivalence class such that $p^{r+1} \dagger k$. So by the above equivalence relation, it follows that for $g \in G$ and $i = 1, \ldots, k, P_i g = P_j$for some,$j, 1 \leq j \leq k$. Let us define $B = \{g \in G: P_1 g = P_1\}$. If $a, b \in B$ then $P_1 ab = (P_1 a)b = (P_1)b = P_1 \Rightarrow ab \in P$. So P is a subgroup of G.

Let $x \in G$ be such that $x \notin B$. We have a coset $Px \neq B$. Now $x \notin B$ we have $P_1 x = P_j$ for some $j \neq 1$. Now any element in Px can be written as hx where $h \in B$. $P_1 hx = (P_1 h)x = P_1 x = P_j$. So for arbitrary coset Px. In this manner we have a coset for each P_i in particular B or P_1. So we will get k number of cosets in G which gives that $o(G) = ko(B), i.e. ko(B) = mp^t$.

But $p^{r+1} \dagger k$ and $p^{t+r} | mp^t = k = o(B) \Rightarrow p^t | o(B) \Rightarrow o(B) \geq p^t$. Now if $p_1 \in P_1$ then for all $h \in B, p_1 h \in P_1 \Rightarrow P_1$ has at least $o(B)$ distinct elements. Since $o(P_1) = p^t$ we have $o(B) \leq p^t \Rightarrow o(B) = p^t$ and B is the required subgroup of G.

Theorem 3.34 (Sylow's Second Theorem)

If p is a prime such that $p^t \mid o(G)$ but $p^{t+1} \dagger o(G)$, then any two subgroups of G of order $p^t$are conjugate.

Before going to prove the theorem let us define the following equivalence relation. For $x, y \in G, x{\sim}y \; if \; y = axb$ for $a \in A, b \in B$ where A and B are two subgroups of G. The equivalence classes for this relation are the double cosets $A \times B = \{a \times b: a \in A, b \in B\}$.

Lemma 3.2

$$o(A \times B) = \frac{o(A)o(B)}{o(A \cap xBx^{-1})}$$

Proof: Let $A \times B \rightarrow A \times Bx^{-1}$be such that $f(a \times b) = a \times bx^{-1}$, f is one-one and onto. So $o(A \times B) = o(AxBx^{-1})$. Now A and xBx^{-1} are sub-groups, so

$$o(AxBx^{-1}) = \frac{o(A)o(xBx)^{-1}}{o(A \cap xBx^{-1})}$$

But $o(xBx^{-1}) = o(B)$. Hence the result follows.

Proof of Theorem 3.34

Let A and B be two subgroups of G of order p^t. Decompose G into double cosets. So $G = \cup AxBo(AxB) = \frac{o(A)o(B)}{o(A \cup xBx^{-1})}$. If $A \neq xBx^{-1}$ for every $x \in G$ then $o(A \cap xBx^{-1})p^r$ where $r < t$. So $o(AxB) = \frac{o(A)o(B)}{p^r} = \frac{p^{2t}}{p^r} = p^{2t-r}$ where $2t - rt + 1$. This implies $p^{t+1} | \, o(AxB)$ for every x and $o(G) = \sum o(AxB)$. So $p^{t+1}|$o(G) which is a contradiction. So $A = xBx^{-1}$ for some $x \in G$, i.e. A and B are conjugate.

We now introduce the following definition.

Definition

If, for some prime p, p^m is the highest power of p dividing the order of G, then the subgroup of order p^m (which exists up to conjugacy) is called p-Sylow subgroup.

Theorem 3.35 (Sylow's Third Theorem)

The number of p-Sylow subgroups in G is equal to $l + kp$ for the prime p and some nonnegative integer k.

Proof: Let H be a p-Sylow subgroup of G. We have

$$G = \cup HxH \text{ and } o(HxH) = \frac{o(H)^2}{o(H \cap xHx^{-1})}.$$

If $(H \cap xHx^{-1}) \neq H$ then $p^{t+1} \mid o(HxH)$ where $p^t = o(H)$. In other words we can write that if $x \notin N(H)$ then $p^{t+1} | \, o(HxH)$. If $x \in N(H)$ then $HxH = H(Hx) = Hx$. So $o(HxH) = o(H) = p^t$.

So we have,

$$o(G) = \sum_{x\in N(H)} o(HxH) + \sum_{x\notin N(H)} o(HxH)$$
$$= \sum_{x\notin N(H)} o(Hx) + \sum_{x\notin N(H)} o(HxH)$$
$$= oN(H) + \sum_{x\notin N(H)} o(HxH)$$

Now in the second term each term is divisible by p^{t+1}. So $\sum_{n\in NH} o(HxH) = np^{t+1}$

$$\Rightarrow o(G) = o\big(N(H)\big) + np^{t+1}. \text{ So}$$
$$\frac{o(G)}{o(N(H))} = 1 + \frac{np^{t+1}}{o(N(H))}$$

As $o(N(H)) \mid o(G)$ the right hand side should be an integer $\Rightarrow \frac{np^{t+1}}{o(N(H))}$ should be an integer.

Again since $p^{t+1} \dagger o\big(N(H)\big)$. So $\frac{np^{t+1}}{o(N(H))}$ must be divisible by p resulting $\frac{np^{t+1}}{o(N(H))} = kp$. Further we know that $o(G) \,|o(N(H))$ is the number of p-Sylow subgroups in G which we just get as $l + kp$.

3.4.1 Application of Sylow's Theorem

Let $o(G) = pq$, where p and q are distinct primes. So, without loss of generality, we can assume $p < q$. We have, at least one subgroup Q of order q. Since q is prime Q has all the elements except the identity of order q. Let G have n subgroups of order q. If $n > 1$ it should be $\geq q + 1$ (by Sylow's third theorem). Each of them has only the identity element common. So the number of elements of order q in G is $\geq (q+1)(q-1) = q^2 - 1$. Hence the number of elements in G is $\geq (q^2 - 1) + 1 = q^2$. But $q^2 > pq$, which is a contradiction. So G has only one Subgroup of order q. Hence Q is self-conjugate, i.e. normal in G.

If $p \dagger (q - 1)$ then $1 + kp \dagger pq$ for any positive integer k because p is prime to $1 + kp$ and $p \dagger (q - 1)$ i.e. $q \neq 1 \ (mod\, p)$. So G has only one subgroup P, of order p; hence P is normal. So we have P and Q as two subgroups of G of order p and q respectively, both normal, such that $P \cap Q = \{e\}$. So we can have $G = PQ$.

Let $a \in P, b \in Q$. So $aba^{-1}b^{-1} = a(ba^{-1}b^{-1}) \in P$ since P is normal and $(aba^{-1})b^{-1} \in Q$ since Q is normal. So $ab\,a^{-1}b^{-1} \in P \cap Q \Rightarrow aba^{-1}b^{-1} = e \Rightarrow ab = ba$. So G is abelian.

Now we have $(ab)^{pq} = a^{pq}.b^{pq} = (a^p)^q(b^p)^q = e \Rightarrow o(ab) \leq pq$. If $o(ab) \leq pq$ then it is either p or q. But this ab should belong to one of the subgroups P or Q in general, which is not the case. So $o(ab) = pq \Rightarrow G$ can be generated by $ab \Rightarrow G$ is cyclic.

Example 3.10

1. There is exactly one 13 Sylow-subgroup in any group of order 130.
   As 13 | 130 and $13^2 \nmid 130$ we have that there exist 13 Sylow-subgroups in any group G of order 130. Now the number of 13 Sylow-subgroups of G is $1 + 13k$ for $k = 0,1,2,\ldots$ and by Sylow's third theorem this number should divide 130. But for $k = 1,2,3,4,5$ we get the numbers as 14, 27, 40, 53, 66 respectively which do not divide 130. For still higher k the number is greater than 130. So only number of the form 1 + 13 k which divides 130 is 1. Hence there is one and only one 13-Sylow subgroup of G.
2. Discuss the number and nature of the 5-Sylow-subgroups and 7-Sylow Subgroups of a group of order $5^2 7^2$. By Sylow's first theorem we know that the group G of order $5^2.7^2$ contains 5-Sylow subgroups and 7-Sylow subgroups. Now the number of 5-Sylow subgroups of G is $1 + 5k$ for $k = 0,1,2,\ldots$ As this number should divide order of G it should divide 7^2 since $1 + 5k$ is prime to 5^2 for any positive integer k. But $1 + 5k$ divides 7^2only for $k = 0$. So there is only one 5-Sylow subgroup of G. Hence it is self conjugate by Sylow's second theorem, i.e. it is normal. Similarly the number of 7- Sylow subgroups of G is $1 + 7k$ which should divide 5^2 and it is true only for $k = 0$. So G has only one 7-Sylow subgroup which is normal.
3. Find all the Sylow-subgroups of the group Z_{24} (The additive group of modulo 24).
   Here we have two types of Sylow-subgroups with order 2^3 and 3 respectively. As the group is abelian all its subgroups are normal hence self conjugate. So we have only one subgroup of 2-Sylow-subgroup and one of 3-Sylow-subgroup. The 2-Sylow-subgroup of order 2^3 contains the elements 0, 3, 6, 9, 12, 15, 18, 21 and the 3-Sylow-subgroup of order 3 contains the elements 0, 8, 16.
4. Show that if G is a group of order 30 then either its 3-Sylow-subgroup or its 5-Sylow subgroup is normal in G. Further show that there exist a normal subgroup of order 15 of G.

G has 3-Sylow subgroup of order 3 and 5 Sylow subgroup of order 5. The number of 3-Sylow subgroups should be $1 + 3j$ for $j = 0, 1, 2, 3...$ and should divide 30. So it is either 1 or 10. The number of 5-Sylow subgroup be $1 + 5j$ for $j = 0, 1, 2, ...$ and should divide 30. So it is either 1 or 6. If in both the cases the number is one then we have both the subgroups normal and the first part is proved.

Let us assume that there are 10, 3-Sylow subgroups and 6, 5-Sylow subgroups in G. Since the subgroups of order 3 and 5 are themselves cyclic and have trivial intersection only, we conclude that there are $10 \times 2 = 20$ elements in G of order 3 and $6 \times 4 = 24$ elements in G of order 5. But this is not possible as G has only 30 elements. So the number of either 3-Sylow subgroups or 5-Sylow subgroups should be one, hence it is normal. So we can get a 3-Sylow subgroup A and a 5-Sylow subgroup B one of which is normal. Again $A \cap B = \{e\}$. So AB has order 15 and also is a subgroup of G. But as AB has index 2 in G, it is normal. So there exists normal subgroup of order 15 in G.

EXERCISES

1. Prove that a group with the property that $x^2 = e$ for every element x is abelian.
2. Show that a group of order n with $n \leq 4$ is abelian.
3. Show that the group of all 1–1 maps of a nonempty set X onto itself is non-abelian.
4. Let G be a cyclic group of order n and H a cyclic group of order m. If m and n are relatively prime, then show that $G \times H$ is cyclic.
5. If G is an abelian group then for all $a, b \in G$ and all integers n, show that $(ab)^n = a^n b^n$.
6. If G is a group such that $(ab)^2 = a^2 b^2$ for all $a, b \in G$, then prove that G must be abelian.
7. If G is a finite group then show that there exists a positive integer N such that $a^N = e$ for all $a \in G$.
8. Show that if every element of a group G is its own inverse, then G is abelian.
9. If G is a group of even order, then show that it has an element $a \neq e$ such that $a^2 = e$.
10. Let G be a finite set which is closed under an associative product and let both cancellation laws hold in G. Show that G is a group.
11. Let G be a group. If p is a prime number and $p \mid o(G)$ then show that G has an element of order p.

12. Let G be a finite abelian group and let the integer n be relatively prime to $o(G)$. Prove that every $g \in G$ can be written as $g = x^n$ with some $x \in G$.
    [*Hint:* Consider the map $\varphi: G \to G$ defined by $\varphi(y) = y^n$ and show that this mapping is an isomorphism of G onto G].
13. Show that if G is a cyclic group of order n and if d is a positive integer such that $d|n$ then G contains exactly one subgroup of order d and this subgroup is cyclic.
14. Let G be an abelian group and H the subset of elements of G of finite order then show that H is a normal subgroup of G.
15. If $G = S_3, H$ and K are Subgroups of G such that $o(H) = 3$ and $o(K) = 2$, then must HK be a subgroup of G?
16. Show that every subgroup of an abelian group is normal.
17. If G is a group and H a subgroup of index two in G then prove that H is a normal subgroup of G.
18. If a cyclic subgroup H of group G is normal in G, then show that every subgroup of H is normal in G.
19. If G is an abelian group and if N is any subgroup of G, then prove that G/N is abelian.
20. If H is the only subgroup of $o(H)$ in a finite group G, then show that H is a normal subgroup of G.
21. Show that the group of automorphisms of a cyclic group of order 8 is a non-cyclic group of order 4.
22. Verify whether the following mappings are homomorphisms:
    (i) Let G be the group of nonzero real numbers under multiplication and $\varphi: G \to G$ defined by,
    $\varphi(x) = x^2$ for all $x \in G$
    (ii) Let G be as in (i) and $\varphi: G \to G$ defined by
    $\varphi(x) = 2^x$ for all $x \in G$
    (iii) Let G be the group of real numbers under addition and $\varphi: G \to G$ defined by,
    $\varphi(x) = x + 1$ for all $x \in G$
    (iv) Let G be as in (iii) and $\varphi: G \to G$ defined by
    $\varphi(x) = 7x$ for all $x \in G$
    (vi) Let G be any abelian group and $\varphi: G \to G$ defined by
    $\varphi(x) = x^5$ for all $x \in G$

23. Verify whether the following mappings are automorphisms:
    (i) Let G be the group of integers under addition and $\varphi: G \to G$ defined by
    $\varphi(x) = -x$ for all $x \in G$.
    (ii) Let G be the group of positive real numbers under multiplication and $\varphi: G \to G$ defined by
    $\varphi(x) = x^2$ for all $x \in G$.
    (iii) Let G be a cyclic group of order 12 and $\varphi: G \to G$ defined by
    $\varphi(x) = x^2$.

CHAPTER 4

Rings and Fields

4.1 DEFINITIONS AND EXAMPLES

Definition

Let R be a non-empty set. Let there be two binary compositions, called addition and multiplication, defined on it. Then R is called a ring if it is:

(a) an abelian group with respect to addition
(b) a semigroup with respect to multiplication
(c) the two distributive laws hold: for $a, b, c \in R$

$$a(b + c) = ab + ac$$

$$(b + c)a = ba + ca$$

If we write explicitly, then the definition of ring will read as follows: A nonempty set R with two binary operations, addition and multiplication, defined on it is called a ring if for all $a, b, c \in R$:

(a) $a + b \in R$
(b) $(a + b) + c = a + (b + c)$
(c) There is an element $0 \in R$ (called zero element) such that $a + 0 = 0 + a = a$
(d) There is an element $-a \in R$ such that $a + (-a) = (-a) + a = 0$
(e) $a + b = b + a$
(f) $ab \in R$
(g) $a(bc) = (ab)c$
(h) $a(b + c) = ab + ac$; $(b + c)a = ba + ca$

A ring R is said to be a ring with unit element (or a ring with identity) if there is an element $1 \in R$ called the multiplicative identity, such that:

$$a1 = 1a = a; \forall a \in R$$

A ring R is called a commutative ring if the multiplication in R satisfies the commutative property, i.e. if:

$$ab = ba; \ \forall a, b \in R$$

A ring R is called a division ring (or a skew field) if all its nonzero elements form a group under multiplication.

A commutative division ring is called a field. In other words a ring R is called a field if all its nonzero elements form an abelian group under multiplication.

If R is a commutative ring, then $0 \neq a \in R$ is said to be a zero divisor if there exists an element $0 \neq b \in R$ such that $ab = 0$.

A commutative ring is called an integral domain if it has no zero divisors.

4.1.1 Examples of Rings and Fields

Example 4.1

1. The following are rings with respect to usual addition and multiplication
   (a) The singleton set consisting only of the number 0
   (b) The set $\mathbb{Z}$ of all integers
   (c) The set of all even integers
   (d) The set $\mathbb{Q}$ of all rational numbers
   (e) The set $\mathbb{R}$ of all real numbers
   (f) The set $\mathbb{C}$ of all complex numbers

   It should be noted that the set in (b) is an integral domain and the sets in (d), (e) and (f) are in fact fields.
2. Any ring of subsets of a set U is a ring with respect to addition and multiplication defined by $A + B = A \Delta B$ and $AB = A \cap B$. The fact that a ring of sets is actually a ring is the reason for the name ring of sets.
   (a) Let m be a positive integer and $\mathbb{Z}_m = \{0, 1, \ldots, m-1\}$. In $\mathbb{Z}_m$ define a sum $a + b$ and the product ab to be the remainders obtained when their usual sum and product are divided by m. Then $\mathbb{Z}_m$ is a ring with respect to the addition and multiplication defined as above. If m is any positive integer, then $\mathbb{Z}_m$ is not always an integral domain. For example, if $m = 6$, then observe that $2 \cdot 3 = 0$ and also $3 \cdot 4 = 0$.
   (b) The set $\{a + b\sqrt{2} : a, b \in \mathbb{Z}\}$is a ring under usual addition and multiplication of real numbers.
   (c) Let $\mathbb{Z}(i) = \{a + ib: a, b \in \mathbb{Z}\}$. Note that a complex number of the form $a + ib$ where $a, b \in \mathbb{Z}$, is called a Gaussian integer and $\mathbb{Z}(i)$ is called the set of all Gaussian integers. $\mathbb{Z}(i)$ is a ring under the usual

addition and multiplication of complex numbers. More precisely it is an integral domain called the domain if Gaussian integers.

(d) The set $\{a + b\sqrt{2} : a, b \in \mathbb{Q}\}$ is a ring under usual addition and multiplication of real numbers.

(e) Let $\mathbb{R}$ denote the field of all real numbers. Then $\mathbb{R} \times \mathbb{R}$ is a field with respect to addition and multiplication of real numbers defined as follows:

$(a, b) + (c, d) = (a + c, b + d)$;

$(a, b) \cdot (c, d) = (ac - bd, ad + bc)$.

Note that $(0, 0)$ and $(1, 0)$ are the additive and multiplicative identities respectively.

If $(a, b) \neq (0, 0) \in \mathbb{R} \times \mathbb{R}$, then $(a, b)^{-1} = \left(\frac{a}{a^2+b^2}, \frac{-b}{a^2+b^2}\right)$ since $(a, b) \cdot \left(\frac{a}{a^2+b^2}, \frac{-b}{a^2+b^2}\right) = (1, 0)$.

(f) The set $\mathbb{R} \times \mathbb{R}$ is a ring under addition and multiplication defined as follows:

$(a, b) + (c, d) = (a + c, b + d)$;

$(a, b) \cdot (c, d) = (ac, bd)$

It is neither an integral domain nor a field. It is not an integral domain because:

$(1, 0) \cdot (0, 1) = (0, 0)$, but $(1, 0) \neq (0, 0)$ and $(0, 1) \neq (0, 0)$.

(g) The set of all n –rowed square matrices forms a ring under addition and multiplication of matrices. This is not an integral domain. First observe that matrix multiplication is not commutative. For if $n = 2$ then,

$\begin{pmatrix} 1 & 0 \\ 0 & 0 \end{pmatrix}\begin{pmatrix} 0 & 1 \\ 1 & 0 \end{pmatrix} = \begin{pmatrix} 0 & 1 \\ 0 & 0 \end{pmatrix}$ but $\begin{pmatrix} 0 & 1 \\ 1 & 0 \end{pmatrix}\begin{pmatrix} 1 & 0 \\ 0 & 0 \end{pmatrix} = \begin{pmatrix} 0 & 0 \\ 1 & 0 \end{pmatrix}$.

Also, by taking $n = 2$ we observe that,

$\begin{pmatrix} 1 & 1 \\ 2 & 2 \end{pmatrix}\begin{pmatrix} -1 & 1 \\ 1 & -1 \end{pmatrix} = \begin{pmatrix} 0 & 0 \\ 0 & 0 \end{pmatrix}$.

This set is also not a skew field. Because any non-zero square matrix A need not have a multiplicative inverse. This holds if and only if A is non-singular.

(h) Let M be the set of all 2×2 matrices of the form $\begin{pmatrix} \alpha & \beta \\ -\bar{\beta} & \bar{\alpha} \end{pmatrix}$ where α, β are complex numbers and $\bar{\alpha}, \bar{\beta}$ are respectively the complex conjugates of α and β. Then M is skew-field under addition and multiplication of matrices. If $A = \begin{pmatrix} \alpha & \beta \\ -\bar{\beta} & \bar{\alpha} \end{pmatrix}$ is a nonzero element of M and if $D = \alpha\bar{\alpha} + \beta\bar{\beta}$, then the matrix $B = \frac{1}{D}\begin{pmatrix} \alpha & \beta \\ -\bar{\beta} & \bar{\alpha} \end{pmatrix}$ is the inverse of A and belongs to M.

We now introduce a special class of rings which are important and useful in various applications.

Definition

A ring R is said to be a Boolean ring if $a^2 = a$ for every $a \in R$.

Example 4.2

(a) The set $\{0, 1\}$ under usual addition and multiplication.
(b) Any ring of sets under addition and multiplication defined by the symmetric difference and intersection property.

4.1.2 Worked Out Examples

Example 4.3

1. If R is a Boolean ring, then
   (a) $a + a = 0; \forall a \in R$
   (b) $a + b = 0$ implies $a = b$

   For every $a \in R, a + a \in R$. Since R is a Boolean ring we have

   $$\begin{aligned}(a + a) + 0 = (a + a) = (a + a)^2 &= (a + a)(a + a) \\ &= a(a + a) + a(a + a) \\ &= (a^2 + a^2) + (a^2 + a^2) \\ &= (a + a) + (a + a)\end{aligned}$$

   By cancellation law we have $a + a = 0$.

   Let $a, b \in R$ be such that $a + b = 0$. But since R is a Boolean ring $a + a = 0$.

   Therefore, $a + a = 0 = a + b$ and by cancellation law it follows that $a = b$.
2. Prove that every Boolean ring R is commutative. Let R be a Boolean ring and let $a, b \in R$.

We have,

$$\begin{aligned} a+b &= (a+b)^2 = (a+b)(a+b) \\ &= a(a+b) + b(a+b) \\ &= a^2 + ab + ba + b^2 \\ &= (a^2+b^2) + ab + ba \\ &= (a+b) + (ab+ba). \end{aligned}$$

Now by cancellation law it follows that,

$ab + ba = 0$

and hence by the previous example we have,

$ab = ba.$

Thus R is commutative.

4.2 IDEALS AND HOMOMORPHISMS

Definition

Let R be a ring and let S be a nonempty subset of R. S is said to be a subring of R if under the composition in R, S itself forms a ring. It is clear that a subring S of a ring R is a subgroup of the additive group R and that it is a semigroup under multiplication. Therefore we can give an alternative definition for a subring and this can be stated as a theorem as follows.

Theorem 4.1

Let R be a ring and let $\Phi \neq S \subset R. S$ is a subring of R if and only if:

(a) $a, b \in S$ implies $a - b \in S$
(b) $a, b \in S$ implies $ab \in S$

The proof of this theorem is easy and is left to the reader as an exercise.

Example 4.4

(a) Let R be a ring. Then R itself and the singleton set $\{0\}$, where 0 is the additive identity in R are subrings of R. These two are called trivial or improper subrings of R and other subrings, if exist, are called nontrivial or proper.

(b) The ring of all even integers is a subring of the ring $\mathbb{Z}$ of all integers under usual addition and multiplication. Note that $\mathbb{Z}$ is a ring with identity, but the ring of all even integers does not have the identity.

(c) In the ring $\mathbb{Z}_6$ (the ring of all integers mod 6) the set $S = \{0, 2, 4\}$ forms a subring.

(d) Let R be a ring. We define Cent R the centre of R as,

Cent $R = \{c \in R: cx = xc \text{ for all } x \in R\}$.

Then Cent R is a subring of R. To show this let $a, b \in \text{Cent } R$ and let $x \in R$ be arbitrary.

Then $ax = xa, bx = xb$. Now,

$(a - b)x = ax - bx = xa - bx = x(a - b)$

and hence $a - b \in \text{Cent } R$. Also

$(ab)x = a(bx) = a(xb) = (ax)b = (xa)b = x(ab)$

Thus $ab \in Cent\ R$ and this completes the proof.

(e) The intersection of two subrings of a ring R is also a subring of R. Let R_1 and R_2 be any two subrings of R. Since R_1 and R_2 are nonempty subsets of R each containing 0, $R_1 \cap R_2$ is a nonempty subset of R. Let $a, b \in R_1 \cap R_2$. Then $a, b \in R_1$and $a, b \in R_2$. Since R_1 and R_2 are subrings, $a - b, ab$ belong to R_1 and also to R_2. Therefore $a - b \in R_1 \cap R_2$ and also $ab \in R_1 \cap R_2$. This shows that $R_1 \cap R_2$ is a subring of R.

Based on our experience in the study of group theory it is now natural to ask whether we can carry over some analog to rings of the concept of normal subgroup. Once this is achieved, one would hope that this analog would lead to the concept of quotient ring in the same manner as quotient group of a group was constructed with the help of a normal subgroup. We now introduce the concept of an ideal for rings. This is the analog to rings of the concept of normal subgroups and the most natural definition is the following.

Definition

A nonempty subset I of a ring R is said to be an ideal of R if:

1. I is a subgroup of R under addition
2. For every $i \in I$ and $r \in R$,
   $ir \in I$ and $ri \in I$.

This is infact called two sided ideal. If R is a commutative ring then it is enough to say that for $i \in I$ and $r \in R, ri \in I$. In any ring R we define a left or right ideal according as $ri \in I$ or $ir \in I$ for all $i \in I$ and $r \in R$. By an ideal we always mean two-sided ideal.

Every ideal is clearly a subring, but converse may not be true.

Example 4.5

(a) The ring $\mathbb{Z}$ of all integers is a subring of the ring (field), $\mathbb{Q}$ of all rational numbers, but $\mathbb{Z}$ is not an ideal in $\mathbb{Q}$.

(b) $\mathbb{Q}$ is a subring (sub field) of the field $\mathbb{R}$ of all real numbers, but $\mathbb{Q}$ is not an ideal in $\mathbb{R}$.

(c) The ring of all even integers is an ideal in the ring $\mathbb{Z}$ of all integers.

(d) Let M be the set of all 2×2 matrices with real entries. Then this is a ring with matrix addition and multiplication. The set of all matrices of the form $\begin{pmatrix} a & b \\ 0 & 0 \end{pmatrix}$ is a right ideal in M where as the set of all matrices of the form $\begin{pmatrix} a & 0 \\ b & 0 \end{pmatrix}$ is a left ideal in M.

(e) If I_1 and I_2 are any two ideals of a ring R then $I_1 \cap I_2$ is an ideal in R.

(f) If I_1 and I_2 are any two ideals of a ring R then $I_1 \cup I_2$ may not be an ideal always in R.

Theorem 4.2

Let R be a ring with (multiplicative) identity and I a proper ideal in R. Then no element of I has a multiplicative inverse in R.

Proof: Suppose, to the contrary, that $a \in I$ has an inverse $a^{-1} \in R$. Since I is an ideal we now have,

$$aa^{-1} = 1 \in I. \text{ Let } r \in R.$$

Then $1r = r \in I$ and hence $R \subset I$. Consequently $I = R$ and this contradicts the given hypothesis that I is a proper ideal of R. Therefore the result follows.

We shall now discuss the quotient of a ring with the ideal just we did in case of a group with the help of normal subgroups.

Let R be a ring and I an ideal in R. Let R/I denote the set of all distinct cosets of I in R which is obtained by considering I as a subgroup of Runder addition. In other words, $R/I = \{a + I : a \in R\}$ where $a + I = \{a + i : i \in I\}$. We now define addition and multiplication in R/I in a natural way as follows:

$$(a + I) + (b + I) = (a + b) + I; \ (a + I)(b + I) = (ab) + I.$$

Theorem 4.3

If I is an ideal in a ring R, then R/I is a ring under addition and multiplication defined as above. If R is commutative, then so is R/I and if R has a unit element, R/I has also a unit element.

Proof: By the result of Chapter 3 it follows that the addition defined above is meaningful and that R/I is group under addition. We shall now show that the multiplication defined above is meaningful. We shall, infact, show that if $a + I = a' + I$ and $b + I = b' + I$ then $(a + I)(b + I) = (a' + I)(b' + I)$, equivalently, we have to show that $ab + I = a'b' + I$. Note that $a + I = a' + I$, therefore $a = a' + i_1$ for some $i_1 \in I$. Similarly $b + I = b' + I$, therefore $b = b' + i_2$ for some $i_2 \in I$. Now $ab = (a' + i_1)(b' + i_2) = a'b' + a'i_2 + i_1b' + i_1i_2$.

Since I is an ideal, $a'i_2, i_1b'$andi_1i_2 are all in I. Consequently, $a'i_2 + i_1b' + i_1i_2 \in I$. Then $ab = a'b' + i_3$ with $i_3 \in I$. From this we conclude that $ab + I = a'b' + I$. We leave it as an exercise for the reader to check that R/I is a ring. Further note that if R is commutative, then so ia R/I for $(a + I)(b + I) = ab + I = ba + I = (b + I)(a + I)$. Also note that if R has unit element 1, then R/I has also unit element $1 + I$.

Theorem 4.4

Let R be a commutative ring with unit element 1 whose only ideals are $\{0\}$and R. Then R is a field.

Proof: We have only to show that for any $0 \neq a \in R$ there exists an element $0 \neq b \in R$ such that $ab = 1$. For this let $0 \neq a \in R$. Define the set $Ra = \{xa: x \in R\}$. Then Ra is an ideal in R. To show that, let $u, v \in Ra$ then $u = r_1a, v = r_2a$ for some $r_1, r_2 \in R$. Thus $u - v = r_1a - r_2a = (r_1 - r_2)$ $a \in Ra$ and hence Ra is a subgroup of R under addition. Further, if $r \in R$ then $ru = r(r_1a) = (rr_1)a \in Ra$. Since R is commutative $ur = ru \in Ra$. This shows that Ra is an ideal in R.

Now by the hypothesis of the theorem, $Ra = \{0\}$ or $Ra = R$. Since $0 \neq a = 1a \in Ra, Ra \neq \{0\}$. Thus $Ra = R$. This means that every element of R is a multiple of a by some element of R. In particular, $1 \in R$ and hence it can be realized as a multiple of a. This means that there exists an element $b \in R$ such that $ba = 1$. This completes the proof of the theorem.

We shall now introduce two special types of ideals in ring and characterize them with the help of their quotient rings.

Definition

Let R be a ring. An ideal P is called *prime ideal* in R if $a, b \in R, ab \in P$ implies either $a \in P$ or $b \in P$. An ideal $M \neq R$ in R is said to be *maximal ideal* of R if whenever I is an ideal in R such that $M \subset I \subset R$ then either $I = R$ or $I = M$.

Theorem 4.5

If R is a commutative ring with unit element and P is an ideal in R, then P is a prime ideal if and only if R/P is an integral domain.

Proof: Let P be a prime ideal in R. First of all note that since R is a commutative ring with unit element, R/P is a commutative ring with unit element. Now let $a, b, \in R$ be such that $ab \in P$. Since P is a prime ideal it follows that either $a \in P$ or $b \in P$. This means that either $a + P = P$ or $b + P = P$. This shows that R/P has no nonzero zero divisor and hence it is an integral domain.

Conversely, let R/P be an integral domain. Let $a, b \in R$ be such that $ab \in P$. That is $(a + P)(b + P) = ab + P = P$. Since R/P has no nonzero zero divisor, either $a + P = P$ or $b + P = P$. That is either $a \in P$ or $b \in P$. Therefore P is a prime ideal and this completes the proof.

Theorem 4.6

If R is a commutative ring with unit element and M is an ideal in R, then M is a maximal ideal if and only if R/M is a field.

Proof: Let M be a maximal ideal in R. Since R is a commutative ring with unit element, R/M is also a commutative ring with unit element. Note that there is a 1–1 correspondence between the set of all ideals of R/M and the set of all ideals of R which contain M. Therefore R/M has only $\{0\}$ and itself as ideals. Thus R/M is a field.

Conversely, let M be an ideal in R and R/M a field. Since R/M is a field its only ideals are $\{0\}$ and R/M itself. But by the correspondence mentioned in the above paragraph, the ideal M of R corresponds to the ideal $\{0\}$ of R/M and R itself corresponds to R/M. Thus there is no ideal in R between M and other than these two. Therefore M is a maximal ideal and this completes the proof.

We have observed in the last chapter that homomorphism of group is an important and useful concept and plays an important role in the study of group theory. This suggests that the appropriate analog for rings could also lead to important ideas. With this motivation we now introduce the concept of homomorphism for rings. Since rings are equipped with two binary compositions a natural definition of homomorphism for rings would be a mapping which preserves both the compositions.

Definitions

Let R and $\bar{R}$ be any two rings. A mapping $f: R \to \bar{R}$ is said to be homomorphism if for all $a, b \in R$:

(i) $f(a+b) = f(a) + f(b)$
(ii) $f(ab) = f(a)f(b)$

A homomorphism of R into $\bar{R}$ is said to be an isomorphism if it is a 1–1 mapping. Two rings R and $\bar{R}$ are said to be isomorphic if there is an isomorphism of R onto $\bar{R}$.

If f is a homomorphism of a ring R into a ring $\bar{R}$ then the kernel of f, denoted by K_f is the set

$$K_f = \{a \in R: f(a) = \bar{0}, \text{the zero of} \bar{R}\}$$

Remark

(a) It should be noted that in the definition of homomorphism addition and multiplication occurring on the left side of (i) and (ii) are those of R and those occurring on the right are those of $\bar{R}$. We used the same notation for convenience and this should not bring any confusion.

(b) If we ignore the multiplication in both the rings R and R then a (ring) homomorphism from R into $\bar{R}$ is at least a homomorphism from R into $\bar{R}$ when we consider them as abelian groups under their respective additions. Therefore as far as addition is concerned, all the properties about homomorphism of groups are carried over to rings. In particular, we have the following result whose proof is left to the reader as an exercise.

Theorem 4.7

If f is a homomorphism of a ring R into a ring $\bar{R}$ then:

(i) $f(0) = \bar{0}$
(ii) $f(-a) = -f(a)$ for each $a \in R$.

Example 4.6

(a) Let R and $\bar{R}$ be any two rings. Define $f: R \to \bar{R}$ by $f(a) = \bar{0}$ for all $a \in R$. Then f is a homomorphism with $K_f = R$.

(b) Let R be any ring. Define $f: R \to R$ by $f(a) = a$ for all $a \in R$. Then f is a homomorphism with $K_f = \{0\}$.

(c) Let R be any ring and S a subring of R. The inclusion map $i: S \to R$ defined by $i(a) = a$ for all $a \in S$ is a homomorphism.

(d) Let R be the ring of integers. Then, $f: R \to R$ defined by $f(x) = 2x$ for all $x \in R$ is not a (ring) homomorphism, although it is group homomorphism, since $f(ab) = 2ab \neq 2a\,2b = f(a)f(b)$.

(e) Let $\mathbb{Z}(\sqrt{2}) = \{a + b\sqrt{2} : a, b \in \mathbb{Z}\}$. Then $\mathbb{Z}(\sqrt{2})$is a ring under usual addition and multiplication of real numbers. Define $f: \mathbb{Z}(\sqrt{2}) \to \mathbb{Z}(\sqrt{2})$ by $f(a + b\sqrt{2}) = a - b\sqrt{2}$. Then f is a homomorphism of $\mathbb{Z}(\sqrt{2})$ into $\mathbb{Z}(\sqrt{2})$ and its kernel $K_f = \{0\}$.

(f) Let $\mathbb{Z}(i) = \{a + bi : a, b \in \mathbb{Z}\}$. Then $\mathbb{Z}(i)$ forms a ring under usual addition and multiplication of complex numbers. Define $f: \mathbb{Z}(i) \to \mathbb{Z}(i)$ by $f(a + bi) = a - bi$. Then f is a homomorphism.

(g) Let $\mathbb{C}$ be the ring (field) of complex numbers. Define $f: \mathbb{C} \to \mathbb{C}$ by $f(z) = \bar{z}$. Then f is a homomorphism.

(h) Let $\mathbb{Z}$ be the ring of integers and $\mathbb{Z}_n$, the ring of integers modulo n. Define $f: \mathbb{Z} \to \mathbb{Z}_n$ by $f(a) =$ remainder of a on division by n. Then f is a homomorphism from $\mathbb{Z}$ onto $\mathbb{Z}_n$ and K_f consists of all multiples of n.

(i) Let $\mathbb{C}$ be the field of complex numbers and M_2 the ring of 2×2 matrices of real entries. Then the map $f: \mathbb{C} \to M_2$ defined by $f(a + bi) = \begin{pmatrix} a & b \\ -b & a \end{pmatrix}$ is an isomorphism.

Theorem 4.8

Let f be a homomorphism from a ring R into a ring $\bar{R}$ with kernel K_f. Then K_f is an ideal in R.

Proof: By results of Chapter 3 it follows that K_f is a normal subgroup of R. Again for all $r \in R$ and $a \in K_f$ we have $f(ra) = f(r)f(a) = f(r)\bar{0} = \bar{0}$, which implies that $ra \in K_f$ and this completes the proof.

Theorem 4.9

Let f be a homomorphism from R into $\bar{R}$. Then f is an isomorphism if and only if $K_f = \{0\}$.

Proof: If f is an isomorphism, clearly $K_f = \{0\}$. Conversely let f be a homomorphism and $K_f = \{0\}$. Suppose $a, b \in R, a \neq b, f(a) = c, f(b) = d$. It is now enough to show that $c \neq d$. Now since $a \neq b, a - b \neq 0$ and hence $a - b \notin K_f$ since $K_f = \{0\}$. In other words $f(a - b) \neq 0$. But f is a homomorphism, so $f(a - b) = f(a) - f(b) = c - d \neq 0$ and hence $c \neq d$. This completes the proof.

We have the following homomorphism theorem for rings and the proof is exactly same as that for groups.

Theorem 4.10

Let R and $\bar{R}$ be rings and f a homomorphism from R into $\bar{R}$ with kernel I. Then $\bar{R}$ is isomorphic to R/I. Moreover, there is a 1–1 correspondence between the set of all ideals of $\bar{R}$ and the set of ideals of R which contain I. This correspondence is achieved by associating an ideal $\bar{I}$ in $\bar{R}$ with ideal I in R defined by $I = \{x \in R: f(x) \in \bar{I}\}$. With I so defined, R/I is isomorphic to $\bar{R}/\bar{I}$.

4.3 INTEGRAL DOMAIN AND FIELD OF QUOTIENTS

Theorem 4.11

Every field is an integral domain, but the converse is not necessarily true.

Proof: Since a field is a commutative ring with identity, in order to prove it is an integral domain it is enough to show that it does not have a nonzero zero divisor. Let F be afield and $a, b \in F$ be such that $ab = 0$. Let $a \neq 0$. Then $a^{-1} \in F$ and hence $ab = 0$ implies that $0 = a^{-1}0 = a^{-1}(ab) = (a^{-1}a)\,b = 1b = b$. Thus $b = 0$ and this completes the proof.

To show that the converse is not always true, consider the ring of all integers under addition and multiplication. This is an integral domain. But it is not a field.

Theorem 4.12

Every finite integral domain is a field.

Proof: Let $D = \{a_1, a_2, \dots, a_n\}$ be a finite integral domain. In order to prove that D is a field we have to:

(i) Produce an element $1 \in D$ such that $1a = a$ for every $a \in D$ and
(ii) For every element $0 \neq a \in D$ produce an element $b \in D$ such that $ab = 1$.

For this let $0 \neq a \in D$. Then the elements $aa_1, aa_2, \dots, aa_n$ are all in D and they all are distinct. For if, $aa_i = aa_j, i \neq j$, then $a(a_i - a_j) = 0$. Since D is an integral domain and $a \neq 0$, we must have $a_i - a_j = 0$ and hence $a_i = a_j$ which contradicts that $i \neq j$. Thus $aa_1, aa_2, \dots, aa_n$ are all distinct elements of D which has exactly n elements. Hence they are precisely all the elements of D, in other words every element $a \in D$ is of the form $a = aa_{i_0}$ for some $a_{i_0} \in D$. Since D is commutative $a = aa_{i_0} = a_{i_0}a$. Now if $y \in d$ then we have $y = aa_i$ for some $a_i \in D$ and so $ya_{i_0} = (aa_i)a_{i_0} = (a_i a)a_{i_0} = a_i(aa_{i_0}) = a_i a = y$. Thus a_{i_0} is a unit element for D and we write it as 1. Now $1 \in D$, and therefore there exists a $b \in D$ such that $ab = 1$. This completes the proof.

Theorem 4.13

If p is a prime number, then $\mathbb{Z}_p$, the ring of integers mod p is a field.

Proof: We shall, infact, show that $\mathbb{Z}_p$is an integral domain. Since it is a finite set, it would then follow from the previous theorem, that $\mathbb{Z}_p$is a field.

Let $i, j \in \mathbb{Z}_p$ and $ij = 0$. Then p must divide ij. Since p is prime, it must divide i or j. Then either $i \equiv 0 (\bmod p)$ or $j \equiv 0 (\bmod p)$, i.e. either $\bar{\imath} = \bar{0}$ or $\bar{\jmath} = \bar{0}$. Thus $\mathbb{Z}_p$ is an integral domain and this completes the proof.

Although, as observed earlier, any integral domain is not a field, but given any integral domain D, we can always construct a field F which will contain properly an isomorphic image of D. Such a field F is called the field of quotients or the quotient field of the integral domain D and the integral domain D is said to be imbedded in the field F. In particular, if we start the construction with the ring of integers, then its quotient field we get is the field $\mathbb{Q}$ of all rational numbers. We now describe the construction.

Theorem 4.14

Every integral domain can be imbedded in a field.

Proof: Let D be the given integral domain. Let M be the set of all ordered pairs (a, b) where $a, b \in D$ and $b \neq 0$. In M define a relation $\sim$ as follows $(a, b) \sim (c, d)$ if and only if $ad = bc$. This defines an equivalence relation on M.

1. If $(a, b) \in M$ then $(a, b) \sim (a, b)$ since $ab = ba$ and hence $\sim$ is reflexive.
2. If $(a, b), (c, d) \in M$ and $(a, b) \sim (c, d)$, then $ad = bc$ hence $cb = da$ and so $(c, d) \sim (a, b)$. This shows that $\sim$ is symmetric.
3. If $(a, b), (c, d), (e, f) \in M, (a, b) \sim (c, d)$ and $(c, d) \sim (e, f)$, then $ad = bc$ and $cf = de$ hence $af = be$. Thus $(a, b) \sim (e, f)$ and the relation $\sim$ is transitive.

Hence $\sim$ divides M into mutually disjoint equivalence classes. Let $[a, b]$ be the equivalence class of (a, b) in M and let F be the set of all such equivalence classes $[a, b]$ where $a, b \in D$ and $b \neq 0$. This F is the field we are seeking. [In particular a rational number is an equivalence class $[a, b]$ of (a, b) where $a, b \in \mathbb{Z}$ and $b \neq 0$. One can also write a/b for rational number $[a, b]$ if there is no confusion. One should note that as order pairs $(1, 2)$ and $(2, 4)$ or as fractions $\frac{1}{2}$ and $\frac{2}{4}$ are distinct, but as rational numbers both are identical because both of them are in the same equivalence class.] It remains to show that the set F defined as above is a field. For this purpose we define addition and multiplication in F. We first dispose of the addition:

$$[a, b] + [c, d] = [ad + bc, bd].$$

Since $b \neq 0$ and $d \neq 0$, it follows that $bd \neq 0$ and hence $[ad + bc, bd] \in F$. We now assert that this addition is well defined, i.e. if $[a, b] = [a', b']$ and $[c, d] = [c', d']$ then $[a, b] + [c, d] = [a', b'] + [c', d']$. Since $[a, b] = [a', b']$ and $[c, d] = [c', d']$ we have $ab' = ba'$ and $cd' = dc'$. Now we have to show $[ad + bc, bd] = [a'd' + b'c', b'd']$. Now,

$$\begin{aligned}(ad + bc)b'd' &= adb'd' + bcb'd' \\ &= ab'dd' + bb'cd' \\ &= ba'dd' + bb'dc' \\ &= bd(a'd' + b'c').\end{aligned}$$

And this is the desired equality. Clearly $[0, b]$ acts as the zero element for this addition, $[-a, b]$ as the additive inverse of $[a, b]$ and F is a commutative group under this addition.

We now turn to the multiplication in F. We define

$$[a, b][c, d] = [ac, bd].$$

Since $b \neq 0$ and $d \neq 0$, it follows that $bd \neq 0$ and hence $[ac, bd] \in F$. In a manner very much similar to the case of addition one can show that if $[a, b] = [a', b']$ and $[c, d] = [c', d']$ then $[a, b][c, d] = [a', b'][c', d']$. One can show that the nonzero elements of F form a commutative group under multiplication where $[d, d]$ acts as the unit element and $[d, c]$ acts as the inverse of $[c, d]$(note that since $c \neq 0, [d, c] \in F$). Also it is a routine verification that the distributive law holds in F and having established all these, we conclude that F is field.

All that remains to show that there is an isomorphism D into F. First notice that for $x \neq 0, y \neq 0$ in $D, [ax, x] = [ay, y]$ since $(ax)y = x(ay)$. Let us denote $[ax, x]$ by $[a, 1]$. Define $f: D \to F$ by $f(a) = [a, 1]$ for every $a \in D$. Clearly f is 1-1 and since:

$$f(a + b) = [a + b, 1] = [a, 1] + [b, 1] = f(a) + f(b)$$

$$f(ab) = [ab, 1] = [a, 1][b, 1] = f(a)f(b)$$

f is a homomorphism. Thus f is an isomorphism from D into F and this completes the proof.

4.4 EUCLIDEAN DOMAIN

In this section we shall discuss a class of rings, in fact a class of integral domain known as Euclidean rings or Euclidean domain which share most of the important properties of the ring of integers under addition and multiplication. The study of such rings has also been motivated by several other well known existing examples—the ring of Gaussian integers, the Jacobian integers and polynomial rings. We shall also discuss the Gaussian integers and the Jacobian integers in this section along with the general concept of a Euclidean ring, whereas the polynomial ring will be discussed separately in another chapter.

Other related concepts namely the concept of principal ideal domain (PID in short) and that of a unique factorization domain (UFD) will also be discussed in this section, we shall also come back to these concepts while dealing with the polynomial rings.

We shall start with the general definition.

Definition

Let R be an integral domain. R is said to be a Euclidean domain if there exists a function d from nonzero elements of R into the nonnegative integers which has the following properties:

(i) for all $a, b \in R, a \neq 0, b \neq 0, d(a) \leq d(ab)$

(ii) for any $a, b \in R, a \neq 0, b \neq 0$, there exist $t, r \in R$ such that $a = bt + r$ where either $r = 0$ or $d(r) < d(b)$.

We do not define $d(0)$. We call $d(a)$ the norm of an element a in a Euclidean ring.

Example 4.7

(a) The ring $\mathbb{Z}$ of integers is a Euclidean domain with the function d defined by $d(a)$ =absolute value of a.

(b) Let $\mathbb{Z}(i)$ denote the set of all complex numbers of the form $a + ib$ where a and b are integers. Then $\mathbb{Z}(i)$ is a Euclidean domain with d defined by $d(a + ib) = a^2 + b^2$.

First of all observe that under usual addition and multiplication of complex numbers $\mathbb{Z}(i)$ is an integral domain, called the domain of Gaussian integers. In order to exhibit that $\mathbb{Z}(i)$ is a Euclidean domain we define a function d on nonzero elements of $\mathbb{Z}(i)$ as follows.

$$d(x) = d(a + ib) = a^2 + b^2$$

where, $x = a + ib \in \mathbb{Z}(i)$ with either $a \neq 0$ or $b \neq 0$. Then clearly $d(x) \geq 1$ i.e. d is a function from nonzero elements of $\mathbb{Z}(i)$ into positive integers. Clearly for any two $x, y \in \mathbb{Z}(i)$ we have,

$$d(xy) = d(x)d(y).$$

This is in fact true for any two complex numbers x and y. Now if $x, y \in \mathbb{Z}(i)$ and $y \neq 0$, then since $d(y) \geq 1$ it follows that,

$$d(x) = d(x)1 \leq d(x)d(y) = d(xy).$$

All that remains is to show that for given $x, y \in \mathbb{Z}(i)$ there exist $t, r \in \mathbb{Z}(i)$ such that $y = tx + r$ where $r = 0$ or $d(r) < d(x)$.

Observe that if we identify an integer $n \in \mathbb{Z}$ with the Gaussian integer $n + i0$, then $\mathbb{Z}$ is a subset of $\mathbb{Z}(i)$. In other words if we define a map: $f: \mathbb{Z} \to \mathbb{Z}(i)$ by,

$$f(n) = n + i0$$

then f is an isomorphism from $\mathbb{Z}$ into $\mathbb{Z}(i)$ and therefore up to isomorphism $\mathbb{Z} \subset \mathbb{Z}(i)$.

We shall first establish the result for a special case, namely the case when $y \in \mathbb{Z}(i)$ is arbitrary but $x = n \in \mathbb{Z}$. Let $y = a + ib$. Then $a, b \in \mathbb{Z}$. By the division algorithm for integers we can now choose u and v such that $a = un + u_1$ and $b = vn + v_1$ where u_1 and v_1 are integers satisfying $|u_1| < \frac{1}{2}n$ and $|v_1| < \frac{1}{2}n$. Let $t = u + iv$ and $r = u_1 + iv_1$. Then,

$$y = a + ib = un + u_1 + i(vn + v_1)$$
$$= (u + iv)n + u_1 + iv_1 = tn + r.$$

Now,

$$d(r) = d(u_1 + iv_1) = u_1^2 + v_1^2 \leq \frac{n^2}{4} + \frac{n^2}{4} = \frac{n^2}{2} < n^2 = d(n)$$

it follows that, in this special case, $y = tn + r$ with $r = 0$ or $d(r) < d(n)$.

We now prove for the general case. Let $x \neq 0$ and let y be an arbitrary element of $\mathbb{Z}(i)$. Then $x\bar{x}$ is a positive integer n where $\bar{x}$ denotes the complex conjugate of x. By applying the result of the particular case to the element $y\bar{x}$ and $n = x\bar{x}$ we observe that there exist elements $t, r \in \mathbb{Z}(i)$ such that $y\bar{x} = tn + r$ with $r = 0$ or $d(r) < d(n)$, i.e.

$$d(r) = d(y\bar{x} - tx\bar{x}) < d(n) = d(x\bar{x}).$$

But

$$d(y\bar{x} - tx\bar{x}) = d\big((y - tx)\bar{x}\big) = d(y - tx)d(\bar{x}).$$

Therefore, $d(y - tx)d(\bar{x}) < d(x\bar{x}) = d(x)d(\bar{x})$. Since $x \neq 0, d(\bar{x}) \neq 0$ and hence we have,

$$d(y - tx) < d(x).$$

Thus we can express y as $y = tx + r_0$ where $r_0 = y - tx, r_0 \in \mathbb{Z}(i)$ and either $r_0 = 0$ or $d(r_0) = d(y - tx) < d(x)$. This proves that $\mathbb{Z}(i)$ is a Euclidean domain.

(c) Let $\mathbb{Z}(\omega) = \{a + \omega b : a, b \in \mathbb{Z}$ and ω is cube root of $1\}$. Note that under usual addition and multiplication $\mathbb{Z}(\omega)$ is an integral domain called the domain of Jacobian integers. Moreover $\mathbb{Z}(\omega)$ is a Euclidean domain. To prove this we define a function d as follows.

For $x = a + \omega b \in \mathbb{Z}(\omega)$, where at least one of a or b is nonzero and ω^2 denotes another cube root of 1, we define,

$$d(x) = (a + \omega b)(a + \omega^2 b) = a^2 - ab + b^2.$$

Since, $a^2 - ab + b^2 = \left(a - \frac{b}{2}\right)^2 + \frac{3}{4}b^2$, if $x \neq 0$, clearly $d(x)$ is a positive integer. Now let $x = a + \omega b$ and $y = c + \omega d$ be any two elements of $\mathbb{Z}(\omega)$. Then since $1 + \omega + \omega^2 = 0$, we have,

$$\begin{aligned} d(x)d(y) &= (a^2 - ab + b^2)(c^2 - cd + d^2) \\ &= \left[\left(a - \frac{b}{2}\right)^2 + \frac{3}{4}b^2\right]\left[\left(c - \frac{d}{2}\right)^2 + \frac{3}{4}d^2\right] \\ &= \left[\left(a - \frac{b}{2}\right) + i\sqrt{\frac{3}{4}}b\right]\left[\left(a - \frac{b}{2}\right) - i\sqrt{\frac{3}{4}}b\right] \\ &\quad \left[\left(c - \frac{d}{2}\right) + i\sqrt{\frac{3}{4}}d\right]\left[\left(c - \frac{d}{2}\right) - i\sqrt{\frac{3}{4}}d\right] \\ &= \left[(ac - bd) - \frac{1}{2}(bc + da) + \frac{1}{2}bd\right]^2 \\ &\quad + \frac{3}{4}[(bc + da) - bd]^2 \\ &= \left[(ac - bd) - \frac{1}{2}(bc + da - bd)\right]^2 \\ &\quad + \frac{3}{4}[(ad + cd +) - bd]^2. \end{aligned}$$

But, $xy = (a + \omega b)(c + \omega d) = (ac - bd) + \omega(bc + ad - bd)$.

Therefore, it follows that $d(xy) = d(x)d(y)$. Now, if $y \neq 0$, since $d(y) \geq 1$ we have

$$d(x) = d(x)1 \leq d(x)d(y) = d(xy).$$

In order to complete the proof we have to show that if x and y are two nonzero elements of $\mathbb{Z}(\omega)$ we can choose two elements $k, r \in \mathbb{Z}(\omega)$ such that $x = ky + r$ where either $r = 0$ or $d(r) < d(y)$. Let $x = a + \omega b$ and $y = c + \omega d$. Then we have,

$$\begin{aligned} \frac{x}{y} = \frac{a + \omega b}{c + \omega d} &= \frac{(a + \omega b)(c + \omega^2 d)}{(c + \omega d)(c + \omega^2 d)} \\ &= \frac{ac + bd - ad}{c^2 - cd + d^2} + \omega\frac{bc - ad}{c^2 - cd + d^2} = R + \omega S \text{ (say)}. \end{aligned}$$

We can choose two integers x_1 and y_1 such that:

$$-\frac{1}{2} \leq R - x_1 \leq \frac{1}{2} \text{ and } -\frac{1}{2} \leq S - y_1 \leq \frac{1}{2}.$$

Now take $k = x_1 + \omega y_1$. Then we get,

$$\begin{aligned} d\left(\frac{x}{y} - (x_1 + \omega y_1)\right) &= (R - x_1)^2 - (R - x_1)(R - y_1) \\ &\quad + (S - y_1)^2 \leq \frac{3}{4} < 1. \end{aligned}$$

Now if we put $r = x - ky$, we get,

$$d(r) = d(x - ky) \leq \frac{3}{4}d(y) < d(y)$$

and this proves the result.

We shall now introduce the definition of a PID and then show that every Euclidean domain is a PID.

Definition

Let R be a commutative ring and let $a \in R$. We write (a) to denote the ideal generated by a or the ideal of all multiples of a. More precisely,

$$(a) = \{xa : x \in R\}.$$

An integral domain R with unit element is said to be principal ideal domain (PID) if every ideal A in R is of the form $A = (a)$ for some $a \in R$.

Theorem 4.15

Every Euclidean domain is a PID.

Proof: Let R be a Euclidean domain and let A be any ideal in R. We shall first show that A is of the form $A = (a)$ for some $a \in R$. If $A = \{0\}$, then by taking $a = 0$ we are through. Now assume that $A \neq \{0\}$. Then there exists an element $0 \neq a \in A$. Choose an $a_0 \in A$ such that $d(a_0)$ is minimal. Note that d takes on values from the set of all nonnegative integers and hence such a choice is possible. Suppose $a \in A$. Since R is a Euclidean domain, there exist $t, r \in R$ such that $a = ta_0 + r$ where $r = 0$ or $d(r) < d(a_0)$. Since $a_0 \in A$ and A is an ideal, $ta_0 \in A$. Therefore $a - ta_0 = r \in A$. Now if $r \neq 0$, then $d(r) < d(a_0)$. This contradicts the choice of a_0. Therefore $r = 0$ and $a = ta_0$. This proves our assertion that A is of the form $A = (a)$for some $a \in R$. We shall now show that every Euclidean domain possesses a unit element. Note that any Euclidean domain R is an ideal of itself. Therefore $R = (a_0)$ for some $a_0 \in R$, i.e. every element in R is a multiple of a_0. Therefore in particular $a_0 = a_0c$ for some $c \in R$. If $a \in A$ then $a = xa_o$ for some $x \in R$. Hence,

$$ac = (xa_0)c = x(a_0c) = xa_0 = a.$$

Thus, c is the unit element (or the multiplicative identity) of R and this completes the proof.

Definition

Let R be a commutative ring and $0 \neq a, b \in R$. a is said to divide b if there exists some $c \in R$ such that $b = ac$. We write $a|b$ to mean that a divides $b$and $a \dagger b$to represent the fact that a does not divide b.

It should be noted that:

(i) $a|b$ and $b|c$ imply $a|c$ (transitivity)
(ii) $a|b$ and $a|c$ imply $a|(b \pm c)$
(iii) $a|b$ implies $a|bx$ for all $x \in R$

These are left to the reader as exercise.

Definition

Let R be a commutative ring and $a, b \in R$. Then $d \in R$ is said to be the greatest common divisor (GCD in short) of a and b if:

(i) $d|a$ and $d|b$
(ii) If $c|a$ and $c|b$, then $c|d$.

We write $d = (a, b)$ to denote the GCD of a and b.

If $a, b \in R$ then $c \in R$ is said to be the least common multiple (LCM in short) of a and b if:

(i) $a|c$ and $b|c$
(ii) If $a|x$ and b$|x$, then $c|x$.

We write $c = [a, b]$ to denote the LCM of a and b.

Theorem 4.16

Let R be a Euclidean domain. Then any two elements $a, b \in R$ have a GCD d. Moreover $d = \lambda a + \mu b$ for some $\lambda, \mu \in R$.

Proof: Let R be a Euclidean domain and $a, b \in R$. Let,

$$A = \{ra + sb: r, s \subset R\}$$

Then, A is an ideal in R. For if $x, y \in A$ then,

$$x = r_1 a + s_1 b, y = r_2 a + s_2 b, \text{ so}$$
$$x - y = (r_1 - r_2)a + (s_1 - s_2)b \in A.$$

Similarly for any $u \in R$

$$ux = u(r_1 a + s_1 b) = (ur_1)a + (us_1)b \in A.$$

Since R is a Euclidean domain, it is a PID. Because A is an ideal in R there exists an element $d \in A$ such that every element of A is a multiple of d. Again $d \in A$ and every element of A is of the form $ra + sb$, therefore we have $= \lambda a + \mu b$ for some $\lambda, \mu \in R$. R has a unit element 1. Therefore, $a = 1a + 0b \in A$ and $b = 0a + 1b \in A$. Now since a, b are in A, they are also multiples of d. Hence $d|a$ and $d|b$. To show $d = (a, b)$, suppose $c|a$ and $c|b$. Then $c|\lambda a$ and $c|\mu b$, so that c divides $\lambda a + \mu b = d$. This completes the proof.

It should be noted that any two elements in a Euclidean ring has a LCM and the proof of this left to the reader as an exercise.

Definition

Let R be a commutative ring with unit element 1. An element $a \in R$ is said to be unit in R if there exists an element $b \in R$ such that $ab = 1$. In other words a unit in a ring is an element whose inverse exists in the ring. Two elements $a, b \in R$ are said to be associates if there exists some unit element $u \in R$ such that $b = ua$.

Observe that for all $a, b, c \in R$:

1. a is an associate of a
2. If a is an associate of b, then b is also an associate of a
3. If a is an associate of b, and b is also an associate of c, then a is an associate of c.

In other words the relation of being associates is an equivalence relation and the verification is left to the reader as an exercise.

Theorem 4.17

Let R be an integral domain with unit element. Let $a, b \in R$ be such that $a|b$ and $b|a$. Then $a = ub$ for some unit $u \in R$.

Proof: Since $a|b$ and $b|a$, we have $b = xa$ and $a = yb$ for some $x, y \in R$. Thus $b = xa = x(yb) = (xy)b$. Since R is an integral domain, cancellation holds and hence $xy = 1$. Thus y is a unit in R and $a = yb$. This completes the proof.

Theorem 4.18

Let R be a Euclidean domain and $a, b \in R$. If $b \neq 0$ is not a unit in R then $d(a) < d(ab)$.

Proof: Let $A = (a) = \{xa: x \in R\}$. Since R is a Euclidean domain, $d(a) \leq d(xa)$ for $0 \neq x \in R$. Now $ab \in A$ and $b \neq 0$. Therefore $d(a) \leq d(ab)$. Let $d(a) = d(ab)$. We can show that every element of A is a multiple of ab. In particular, since $a \in A, a$ must be a multiple of ab. Therefore $a = abx$ for some $x \in R$. Since R is basically an integral domain, we have $bx = 1$. This contradicts the fact that b is not a unit in R. Thus $d(a) < d(ab)$ and this completes the proof.

Definition

Let R be an integral domain. A non unit $\pi \in R$ is said to be a prime element of R if whenever $\pi = ab$ with $a, b \in R$, then one of a or b is a unit in R. In other words, a prime element is an element which cannot be factored in nontrivial way. Two elements a and b in R are said to be relatively prime if their GCD is a unit of R.

Since any associate of a GCD is a GCD and since 1 is an associate of any unit, if a and b are relatively prime, we may assume that $(a, b) = 1$.

Theorem 4.19

The ideal $A = (a_0)$ is a maximal ideal of the Euclidean domain R if and only if a_0 is a prime element of R.

Proof: First suppose that a_0 is not a prime element. Therefore $a_0 = bc$ for some $b, c \in R$ such that neither b nor c is a unit. Let B be the ideal generated by b, i.e. $B = (b)$. Then, certainly $a_0 \in B$, so that $A \subset B$.

If $B = R$ then $1 \in B$. So that $1 = xb$ for some $x \in R$, this contradicts the fact that b is not a unit in R. If $A = B$ then $b \in B = A$ so that $b = xa_0$ for some $x \in R$. Now, $a_0 = bc = xa_0c = xca_0$. Thus, $xc = 1$. This again contradicts the fact that c is not a unit in R. Therefore $A \subset B, A \neq B, B \neq R$ and thus A cannot be a maximal ideal of R.

Conversely, assume that a_0 is a prime element of R and that U is an ideal of R with the property that $A = (a_0) \subset U \subset R$. Since R is a Euclidean domain, $U = (u_0)$. Since $a_0 \in A \subset U = (u_0), a_0 = xu_0$ for some $x \in R$. But since a_0 is a prime element of R, either x or u_0 must be a unit in R. Observe that if an ideal U of a ring contains a unit, then U must be the whole ring.

Therefore, in this case, if u_0 is a unit, then $U = R$. If x is a unit in R, then $x^{-1} \in R$. Since $a_0 = xu_0$ we have $u_0 = x^{-1}a_0$. Again since A is an ideal, $u_0 \in A$ showing that $U \subset A$. But since $A \subset U$ we have $A = U$. Therefore there exists no ideal of R which lies strictly between A and R and hence A is a maximal ideal of R.

Theorem 4.20

Let R be a Euclidean ring. Suppose that for $a, b, c \in R, a|bc$ but $(a, b) = 1$. Then, $a|c$.

Proof: Since 1 is GCD of a and b, there exist $\lambda, \mu \in R$ such that $\lambda a + \mu b = 1$. Hence, $\lambda ac + \mu bc = c$. Now, $a|\lambda ac$. Since by hypothesis $a|bc, a|\mu bc$. Therefore, $a|(\lambda ac + \mu bc) = c$.

Theorem 4.21

(i) If π is a prime element in a Euclidean ring R and $a \in R$, then either $\pi|a$ or $(\pi, a) = 1$.

(ii) If π is a prime element in a Euclidean ring R and $\pi|a_1a_2$where$a_1, a_2 \in R$, then π divides at least one of $a_1$or a_2. This result can be extended to any finite number of elements $a_1, a_2, \dots, a_n \in R$.

Proof:

(i) Let π be a prime element of R and $a \in R$. Then either $\pi|a$ or $(\pi, a) = 1$, for, in particular (π, a) is a divisor of π, so it must be π or 1 or any unit. If $(\pi, a) = 1$ then we are through. If $(\pi, a) = \pi$, since $(\pi, a)|a$, we get $\pi|a$ and hence the result follows.

(ii) Let π be a prime element of R. Let $\pi|a_1a_2$ for $a_1, a_2 \in R$. Let $\pi \dagger a_1$. Then $(\pi, a_1) = 1$. Hence $\pi|a_2$ and this completes the proof.

We shall now define one more concept for an integral domain, called Unique Factorisation Domain (UFD in short), and show that every Euclidean domain is a UFD.

Definition

Let R be an integral domain with unit element, R is said to be UFD if:

(i) any nonzero element in R is either a unit or can be written as the product of a finite number of prime elements (or irreducible elements) of R.

(ii) the decomposition described in (i) is unique up to the order and associates of the prime elements.

The following two theorems show that every Euclidean domain is a UFD.

Theorem 4.22

Let R be a Euclidean ring. Then every nonzero element in R is either a unit in R or can be written as the product of a finite number of prime elements of R.

Proof: Let 'a' be a nonzero element of R. We have to show that either a is a unit in R or can be written as the product of a finite number of prime elements of R. We shall prove this by induction on $d(a)$.

Note that $d(a) = d(1a) \geq d(1)$.

If $d(a) = d(1)$, then a is a unit element in R and thus the result is true for $d(1)$.

Now assume that the result is true for all nonzero elements $x \in R$ such that $d(x) < d(a)$. We shall now show that the result is true for a.

If a is a prime element of R, the result is obviously true. So suppose that a is not a prime. Then we can write $a = bc$ where neither b nor c is a unit in R. Since both b and c are nonunits, we have

$$d(bc) > d(b), d(bc) > d(c).$$

But, $d(a) = d(bc)$.

Therefore, $d(b) < d(a), d(c) < d(a)$.

So by induction hypothesis each of b *and* c can be written as the product of a finite number of prime elements of R. Let $b = p_1 p_2 \cdots p_n, c = q_1 q_2 \cdots q_m$ where $p_1, p_2, \dots, p_n, q_1, q_2, \dots, q_m$ are primes. So $a = p_1 p_2 \cdots p_n q_1 q_2 \cdots q_m$.

Theorem 4.23

Let R be a Euclidean ring and 'a' a nonzero nonunit in R. Suppose

$$a = p_1 p_2 \cdots p_m = q_1 q_2 \cdots q_n$$

where $p_1, p_2, \dots, p_m, q_1, q_2, \dots, q_n$ are prime elements in R. Then $m = n$ and each p is an associate of some q and each q is an associate of some p.

Proof: Given that:

$$p_1 p_2 \cdots p_m = q_1 q_2 \cdots q_n.$$

Since $p_1 | p_1 p_2 \cdots p_m, p_1 | q_1 q_2 \cdots q_n$. So p_1 must divide at least one of $q_1, q_2, \dots, q_n$. R is a commutative ring, so without any loss of generality, we may assume that $p_1 | q_1$. But p_1 and q_1 are both primes. Therefore p_1 and q_1 must be associates. So $q_1 = u_1 p_1$ where u_1 is a unit in R. thus we have:

$$p_1 p_2 \cdots p_m = u_1 p_1 q_2 \cdots q_n.$$

Since $p_1 \neq 0$ by cancellation we get,

$$p_2 p_3 \cdots p_m = u_1 q_2 q_3 \cdots q_n.$$

Now repeat the above procedure with p_2 and so on. If $n > m$ after m steps we have:

$$1 = u_1 u_2 \cdots u_m q_{m+1} q_{m+2} \cdots q_n.$$

where, $u_1, u_2, \dots, u_m$ are units in R. Since $q_{m+1}, q_{m+2}, \dots, q_n$ are primes they are not units. So the above equation is impossible, the product of some units and some primes cannot be equal to 1. So $n \ngtr m$, i.e $n \leq m$.

Similarly by interchanging the roles of p's and q's we get $m \leq n$.

Thus, $m = n$ and in the process we have also shown that every p is an associate of some q and conversely every q is an associate of some p.

Combining the above two results we obtain that every nonzero element in a Euclidean domain can be uniquely written up to associates and order) as a product of prime elements or is a unit in R. This is in other words, says that every Euclidean domain is a UFD.

We shall complete this section by calculating the units and prime elements of Gaussian domain.

Theorem 4.24

An element of $\mathbb{Z}(i)$ is a unit if and only if its norm is 1.

Proof: Let e be a unit. Then for some η,

$$1 = e\eta.$$

So

$$1 = d(e\eta) = d(e)d(\eta).$$

Hence $d(e)$ divides 1. $d(e)$ being the sum of squares of two integers,

$$d(e) = 1.$$

On the other hand, if $d(a + ib) = 1$ then

$$1 = a^2 + b^2 = (a + ib)(a - ib).$$

So $a + ib$ must divide 1 and hence $a + ib$ is a unit.

Theorem 4.25

$1, -1. i. -i$ are the only units of $\mathbb{Z}(i)$.

Proof: Let $a + ib$ be a unit. Then

$$1 = d(a + ib) = a^2 + b^2$$

a and b being integers, $a^2 = 1$ or $b^2 = 1$. So $a = \pm 1, b = 0$ or $a = 0, b = \pm 1$. So $a + ib$ is one among $1, -1, i$ and $-i$.

If e is a unit and $\xi \in \mathbb{Z}(i)$, then $e\xi$ is an associate of ξ. Therefore $\xi, -\xi, i\xi$ and $-i\xi$ are the only associates of ξ.

It is clear that if ξ divides η and if e_1 and e_2 are units, then $e_1\xi$ divides $e_2\eta$. In fact, any associate of ξ divides any associate of η.

Any nonzero element of $\mathbb{Z}(i)$ which is not a unit is a prime element if it is only divisible by its associates and by units. If π is a prime element of $\mathbb{Z}(i)$, then π is not divisible by any element except $i, \pi, -1, -\pi, -i\pi, -i, i\pi$.

Theorem 4.26

If the norm of a Gaussian integer is a prime number, then it must be a prime Gaussian number.

Proof: Let p be a prime number and $\xi \in \mathbb{Z}(i)$ be such that $d(\xi) = p$. If $\xi = \eta\rho$ for $\eta, \rho \in \mathbb{Z}(i)$, then

$$p = d\xi) = d(\eta)d(\rho).$$

So either $d(\eta) = 1$ or $d(\rho) = 1$. Hence either η or ρ is a unit. So ξ is a prime Gaussian number.

Since $d(2 + i) = 5, d(4 + i) = 17, d(5 + 2i) = 29, d(6 + 5i) = 61$ and since $5, 17, 29, 61$ are prime numbers, $2 + i, 4 + i, 5 + 2i, 6 + 5i$ are prime Gaussian numbers.

Remarks

A prime number may or may not be a prime Gaussian number. Since,

$$5 = (2 + i)(2 - i)$$
$$17 = (4 + i)(4 - i)$$
$$29 = (5 + 2i)(5 - 2i)$$
$$61 = (6 + 5i)(6 - 5i)$$

$5, 17, 29$ and 61 are not prime Gaussian numbers although they are prime numbers, but 7 is a prime Gaussian number. To prove this let a, b, c, d be integers such that:

$$7 = (a + ib)(c - id)$$

Then, $7 = ac + bd$ and $0 = bc - ad$. Let $\frac{c}{a} = \frac{d}{b} = k$.

Then, $7 = ac + bd = k(a^2 + b^2) = \frac{1}{k}(c^2 + d^2)$, that is k is a positive number. If $k \leq 1$. $7 = \frac{1}{k}(c^2 + d^2) \geq (c^2 + d^2)$.

But since $49 = d(7) = d(a + ib)d(c + id) = (a^2 + b^2)(c^2 + d^2)$, $(c^2 + d^2)$ must divide 49. So $c^2 + d^2 = 7$ or 1. But $7 = 0 + 7 = 1 + 6 = 2 + 5 = 3 + 4$.

Therefore, for $c \neq 0 \neq d, c^2 + d^2 \neq 7$. So $c^2 + d^2 = 1$. Similarly $k \geq 1$, then $a^2 + b^2 = 1$. So either $(c - id)$ or$(a + ib)$ is a unit. Thus 7 is a prime element in $\mathbb{Z}(i)$. Using this fact one can show that.

Theorem 4.27

Any prime number which is not the sum of two squares is a prime in $\mathbb{Z}(i)$.

Note that any prime number p which can be expressed as the sum of two squares cannot be prime in $\mathbb{Z}(i)$ because, if $p = m^2 + n^2$, then $p = (m + in)(m - in)$.

EXERCISES

1. Let R be any ring and $a, b \in R$. Find an expression for $(a + b)^2$ and what is the form of this when R is commutative.
2. Let R be any ring and $a, b \in R$. Find an expression for $(a + b)^n$ where n is a positive integer. What is the form of this when R is commutative?
3. If I is an ideal in a ring R with identity 1 and if $1 \in I$ then prove that $I \neq R$.
4. Let R be a ring with identity, not necessarily commutative, such that the only right ideals of R are (0) and R. Prove that R is a division ring.
5. If R is a ring with identity, show that any divisor of zero in R is singular.
6. Let R be the ring of all real functions defined on $[0, 1]$. If X is a subset of $[0, 1]$. Show that the set $I(X)$ in R defined by,
   $I(X) = \{f: f(x) = 0 \text{ for every } x \in X\}$
   is an ideal. Show that this ideal is maximal if and only if X is singleton.
7. What are the regular elements, singular elements and divisors of zero in the ring of all subsets of a non empty set under symmetric difference and intersection? Under what condition is this ring a field?
8. In the ring of integers whether 2 is regular nor singular or neither and why?
9. Prove that any homomorphism of a field is either an isomorphism or takes each element to zero.

10. Find out the units of the Jacobian domain $\mathbb{Z}(\omega)$.
11. Prove that the set:
    $\{a + b\sqrt{p}: a, b \in \mathbb{Q}, p \text{ is a prime}\}$
    is a field under the usual addition and multiplication.
12. Prove that if a ring R has a left identity and a right identity then both are equal.
13. An element a in a ring R is said to be idempotent if $a^2 = a$. Prove that the only idempotent elements of an integral domain with unity are 0 and 1.
14. An element a in a ring R is said to be nilpotent if $a^n = 0$ for some positive integer n. prove that only nilpotent element of an integral domain is the additive identity 0.
15. Let U is an ideal of a ring R and:
    $r(U) = \{x \in R: xu = 0 \text{ for } u \in U\}$.
    Prove that $r(U)$ is an ideal in R.
16. Let U and V be ideals of a ring R and:
    $U + V = \{u + v: u \in U, v \in V\}$.
    Prove that $U + V$ is also an ideal of R.
17. Let R be a commutative ring with unit element 1. Define a relation '$\sim$' in R as follows
    $a \sim b$ iff a is an associate of b. Prove that this is an equivalence relation.
18. Show that every homomorphic image of a commutative ring is a commutative ring.
19. If R is a ring with unit element 1 and f is homomorphism of R onto R'. Prove that $f(1)$ is the unit element of R'.
20. Let R be a commutative ring with unity. Then prove that every maximal ideal of R is a prime ideal.
21. If R is a finite commutative ring with unity, prove that every prime ideal of R is a maximal ideal.
22. Prove that the ring of integers is a Euclidean domain.
23. Prove that every field is a Euclidean domain.
24. Prove that the necessary and sufficient condition that a nonzero element 'a' in a Euclidean domain is a unit is that: $d(a) = d(1)$.
25. Find the principal ideal generated by 5 in the ring of integers $\mathbb{Z}$.
26. Prove that the quotient field of a finite integral domain coincides with itself.
27. Give a description of the field of quotient of the integral domain:
    $D = \{a + b\sqrt{2}: a, b \in \mathbb{Z}\}$.

CHAPTER 5

Polynomial Rings and Roots of Polynomials

5.1 POLYNOMIAL RINGS

Definition

Let R be a commutative ring with unit element. By a polynomial $p(x)$ over R in the indeterminate x, we mean a symbol of the form $p(x) = a_0 + a_1x + a_2x^2 + ... + a_nx^n$ where n is any nonnegative integer and $a_0, a_1, ..., a_n$ are all in R. $a_0, a_1, ..., a_n$ are called the coefficients of polynomial $p(x)$. We write $R[x]$ to denote the set of all polynomials over R in the indeterminate x. We write $O(x)$ to denote the polynomial whose all the coefficients are zeros and call it the zero polynomial. $1(x)$ denotes the polynomial with $a_0 = 1$ and all other coefficients zeros. Two polynomials $p(x) = a_0 + a_1x + a_2x^2 + ... + a_mx^m$ and $q(x) = b_0 + b_1x + ... + b_nx^n$ are said to be equal and written as $p(x) = q(x)$ if for every nonnegative integer i, $a_i = b_i$.

We shall now define addition and multiplication in the set $R[x]$.

Definition

If $p(x) = a_0 + a_1x + a_2x^2 + ... + a_mx^m$ and $q(x) = b_0 + b_1x + ... + b_nx^n$ are in $R[x]$ then we define their sum as,

$$p(x) + q(x) = c_0 + c_1x + c_2x^2 + ... + a_tx^t$$

Where for each i, $c_i = a_i + b_i$.

The product of $p(x)$ and $q(x)$ is defined by,

$$p(x)q(x) = c_0 + c_1x + ... + c_kx^k$$

Where $c_i = a_ib_0 + a_{i-1}b_1 + a_{i-2}b_2 + ... \ a_0b_i$

$$= \sum_{k=0}^{i} a_kb_{i-k}$$

Example 5.1

(a) Let $p(x) = 1 + x,\ q(x) = 3 - 2x + x^2$. We consider $p(x) = 1 + x + 0.x^2$.

Here $a_0 = 1,\ a_1 = 1,\ a_2 = 0,\ a_3 = a_4 = \ldots = 0$

$b_0 = 3,\ b_1 = -2,\ b_2 = 1,\ b_3 = b_4 = \ldots = 0$

$$p(x) + q(x) = (1+3) + (1-2)x + (0+1)x^2$$
$$= 4 - x + x^2$$

For the multiplication we have,

$c_0 = a_0b_0 = 1.3 = 3$

$c_1 = a_1b_0 + a_0b_1 = 1.3 + 1.(-2) = 1$

$c_2 = a_2b_0 + a_1b_1 + a_0b_2 = 0.3 + 1.(-2) + 1.1 = -2 + 1 = -1$

$c_3 = a_3b_0 + a_2b_1 + a_1b_2 + a_0b_3 = 0.3 + 0.(-2) + 1.1 + 1.0 = 1$

$c_4 = a_4b_0 + a_3b_1 + a_2b_2 + a_1b_3 + a_0b_4 = 0.3 + 0.(-2) + 0.1 + 1.0 + 1.0 = 0$

$c_5 = c_6 = \ldots = 0$

Therefore,

$$p(x)q(x) = 3 + x - x^2 + x^3$$

(b) Let $p(x) = 1 + x - x^2,\ q(x) = 2 + x^2 + x^3$

Here $a_0 = 1,\ a_1 = 1,\ a_2 = -1,\ a_3 = a_4 = \ldots = 0$

$b_0 = 2,\ b_1 = 0,\ b_2 = 1,\ b_3 = 1,\ b_4 = b_5 = \ldots = 0$

$$p(x) + q(x) = (1+2) + (1+0)x + (-1+1)x^2 + (0+1)x^3$$
$$= 3 + x + 0.x^2 + x^3$$
$$= 3 + x + x^3$$

For the product

$c_0 = a_0b_0 = 1.2 = 2$

$c_1 = a_1b_0 + a_0b_1 = 1.2 + 1.0 = 2$

$c_2 = a_2b_0 + a_1b_1 + a_0b_2 = (-1).2 + 1.0 + 1.1 = -2 + 1 = -1$

$c_3 = a_3b_0 + a_2b_1 + a_1b_2 + a_0b_3 = 0.2 + (-1).0 + 1.1 + 1.1 = 2$

$c_4 = a_4b_0 + a_3b_1 + a_2b_2 + a_1b_3 + a_0b_4 = 0.4 + 0.0 + (-1).1 + 1.1 + 1.0 = 0$

$c_5 = a_5b_0 + a_4b_1 + a_3b_2 + a_2b_3 + a_1b_4 + a_0b_5$
$= 0.2 + 0.0 + 0.1 + (-1).1 + 1.0 + 0.0 = -1$

$c_6 = a_6b_0 + a_5b_1 + a_4b_2 + a_3b_3 + a_2b_4 + a_1b_5 + a_0b_6$
$= 0.2 + 0.0 + 0.1 + 0.1 + (-1).0 + 1.0 + 1.0 = 0$

$c_7 = c_8 = \ldots = 0$

Hence,

$$p(x)q(x) = c_0 + c_1x + c_2x^2 + c_3x^3 + c_4x^4 + c_5x^5 + \ldots$$
$$= 2 + 2x - x^2 + 2x^3 - 0.x^4 - x^5 + 0.x^6 + \ldots$$
$$= 2 + 2x - x^2 + 2x^3 - x^5$$

Definition

Let $p(x) = a_0 + a_1x + a_2x^2 + \ldots + a_nx^n$ be a nonzero polynomial in $R[x]$. If $a_n \neq 0$ then the degree of $p(x)$ is defined to be n. In other words, the degree of a polynomial $p(x)$ is the largest integer i for which the ith coefficient a_i of $p(x)$ is not zero.

The degree of the zero polynomial is not defined. A polynomial is said to be a constant polynomial (or simply a constant) if its degree is 0.

Theorem 5.1

Let R be a commutative ring with unit element. Then $R[x]$ the set of all polynomials over R in the indeterminate x is a commutative ring with unit element under addition and multiplication defined as above.

Proof: It follows from the definition that $R[x]$ is closed under addition and multiplication. We leave it to the reader as an exercise to verify that addition and multiplication are associative, commutative and distributive. The zero polynomial and the identity polynomial serve respectively as the additive and multiplicative unit elements. Given any polynomial $p(x) \in R(x)$, $p(x) = a_0 + a_1x + a_2x^2 + \ldots + a_nx^n$ the polynomial $-p(x) = (-a_0) + (-a_1x) + (-a_2)x^2 + \ldots + (-a_n)x^n$ is its additive inverse.

Theorem 5.2

Let R be an Integral domain. If $p(x)$ and $q(x)$ are any two nonzero polynomials in $R[x]$ then $\deg(p(x)q(x)) = \deg p(x) + \deg q(x)$.

Thus $\deg p(x) \leq \deg(p(x)q(x))$ and $\deg q(x) \leq \deg(p(x)q(x))$.

Proof: Let $p(x) = a_0 + a_1x + a_2x^2 + \ldots + a_mx^m$

$$q(x) = b_0 + b_1x + b_2x^2 + \ldots + b_nx^n$$

with $a_m \neq 0$, $b_n \neq 0$.

Therefore, $\deg p(x) = m$, $\deg q(x) = n$ and hence $\deg p(x) + \deg q(x) = m + n$.

By definition

$$p(x)q(x) = c_0 + c_1x + c_2x^2 + ... + c_kx^k$$

where $c_t = \sum_{i=0}^{t} a_i b_{t-i}$

We have to show that $\deg(p(x)q(x)) = m + n$. Therefore it is enough to show that $c_{m+n} = a_m b_n \neq 0$ and $c_i = 0$ for $i > m + n$. But it follows at once from the definition that $c_{m+n} = a_m b_n$. Again $a_m \neq 0$, $b_n \neq 0$ and R is an integral domain. Therefore $a_m b_n \neq 0$. Now c_i is the sum of the terms of the form $a_j b_{i-j}$, since $i = j + (i - j) > m + n$ then either $j > m$ or $(i - j) > n$. But then one of a_j or b_{i-j} is 0. Therefore c_i, being the sum of zeros, is zero and this completes the proof.

Theorem 5.3

If R is an integral domain, then

(i) $R[x]$ is also an integral domain

(ii) $R[x]$ is not a field.

Proof

(i) Let R be an integral domain and let $p(x) \neq 0$ and $q(x) \neq 0$ be in $R[x]$. Since R is an integral domain we have,

$$\deg(p(x)q(x)) = \deg p(x) + \deg q(x)$$

Now it is impossible to have $p(x)q(x) = 0$ and therefore $R[x]$ is an integral domain.

(ii) Let $p(x) \in R[x]$ and $\deg p(x) > 0$. Since degree of the identity polynomial 1 is 0, there exists no $q(x) \in R[x]$ such that $\deg p(x) + \deg q(x) - \deg 1 = 0$. Thus no polynomial of nonzero degree has a multiplicative inverse and thus $R[x]$ is not a field.

Theorem 5.4 (Division Algorithm)

Let F be a field and let $p(x)$ and $q(x) \neq 0$ be any two polynomials in $F[x]$. Then there exist two polynomials $t(x)$ and $r(x)$ in $F[x]$ such that $p(x) = t(x)$ $q(x) + r(x)$.

Where $r(x) = 0$ or $\deg r(x) < \deg q(x)$.

Proof: If $\deg p(x) < \deg q(x)$ then take $t(x) = 0$ and $r(x) = p(x)$. Then we have $p(x) = 0q(x) + p(x)$ where $p(x) = 0$ or $\deg p(x) < \deg q(x)$ and this completes the proof. Now assume that $\deg p(x) \geq \deg q(x)$. More precisely, let $p(x) = a_0 + a_1x + ... + a_mx^m$, $q(x) = b_0 + b_1x + ... + b_nx^n$, $a_m \neq 0$, $b_n \neq 0$, $m \geq n$. Since F is a field and $b_n \in F$, $b_n^{-1} \in F$. Put

$$p_1(x) = p(x) - a_mb_n^{-1}x^{m-n}q(x)$$

Then $\deg p_i(x) \leq m-1$. By mathematical induction on the degree of $p(x)$ we assume that,

$$p_i(x) = t_1(x)q(x) + r(x)$$

where $r(x) = 0$ or $\deg r(x) < \deg q(x)$.

Then $p(x) = a_mb_n^{-1}x^{m-n}q(x) + t_1(x)q(x) + r(x)$.

Put $t(x) = a_mb_n^{-1}x_{m-n} + t_1(x)$.

Then we have

$$p(x) = t(x)q(x) + r(x)$$

where $r(x) = 0$ or $\deg r(x) < \deg q(x)$.

This completes the proof.

As a corollary of this theorem we obtain the following important result.

Theorem 5.5

If F is a field, then $F[x]$ is an Euclidean domain

Proof: Let $p(x)$ be a nonzero polynomial in $F[x]$. Then we define a function d by $d(p(x)) = \deg p(x)$.

Then from the definition it is clear that $d(p(x)) \geq 0$ for each nonzero polynomial $p(x)$.

For all nonzero $p(x), q(x) \in F[x]$ we have $d(p(x)q(x))$. Also it follows from the previous theorem that for all nonzero polynomials $p(x), q(x) \in F[x]$ there exist $t(x), r(x) \in F[x]$, such that $p(x) = t(x)q(x) + r(x)$ where $r(x) = 0$ or $\deg r(x) < \deg q(x)$ and this completes the proof.

Example 5.2

Consider $p(x) = 2x^3 - 4x^2 + 5$ and $q(x) = 2x^2 - 3x + 2$ as two nonzero polynomials over the real field R. We apply the division algorithm as follows

$$2x^3 - 4x^2 + 5 = \left(x - \frac{1}{2}\right)(2x^2 - 3x + 2) + \left(\frac{7}{-2}x + 6\right)$$

That is, $p(x) = t(x)q(x) + r(x)$

where $t(x) = x - \frac{1}{2}$ and $r(x) = -\frac{7}{2}x + 6$

Observe that $\deg r(x) = 1 < \deg q(x) = 2$.

Remark

We have just observed that if F is a field then $F[x]$ is a Euclidean domain. Hence we talk of *gcd*, *lcm*, units, associates and prime elements in $F[x]$. But it should be noted that in $F[x]$ the only units are the nonzero constant polynomials (i.e. the polynomials of degree 0). Similarly the prime elements of $F[x]$ have a different name, they are called irreducible polynomials. For example a polynomial $p(x) \in F[x]$ is said to be irreducible over F if whenever $p(x) = a(x)b(x)$ with $a(x), b(x) \in F[x]$ then one of $a(x)$ or $b(x)$ has degree 0 (i.e. is a constant). Note that irreducibility depends on the field, the polynomial $x^2 + 1$ is irreducible over $\boldsymbol{R}$ but not over $\boldsymbol{C}$. In $\boldsymbol{C}$, $x^2 + 1 = (x + i)(x - i)$. Similarly $x^2 - 2$ is irreducible over $\boldsymbol{Q}$ but not over $\boldsymbol{R}$. We shall come back to this point again when we shall discuss the roots of polynomials.

We now state some important results of Euclidean ring which are, in particular, true for $F[x]$ where F is a field. The reader may try to prove these for this particular case by using the arguments used to prove the general results for Euclidean domain.

Theorem 5.6

(a) $F[x]$ is a PID.

(b) Any two polynomials $p(x), q(x) \in F[x]$ have a *GCD* $d(x)$ and this can be expressed as,

$d(x) = a(x)p(x) + b(x)q(x)$

(c) $F[x]$ is a UFD.

(d) The ideal $A=(p(x))$ in $F[x]$ is a maximal ideal if and only if $p(x)$ is irreducible over F.

We shall now work out some examples.

Example 5.3

(a) If $f(x)=x^3-2x^2+3x-7$ and $g(x)=x^2+2$ then find out $d(x)$ the *GCD* of $f(x)$ and $g(x)$ and express $d(x)$ as,

$d(x)=a(x)f(x)+b(x)g(x)$

By the division algorithm.

$$(x^3-2x^2+3x-7)=(x-2)(x^2+2)+(x-3) \quad \ldots \text{(i)}$$

$$x^2+2=(x+3)(x-3)+11 \quad \ldots \text{(ii)}$$

$$x-3=\left(\frac{x}{11}-\frac{3}{11}\right)11+0 \quad \ldots \text{(iii)}$$

The constant polynomial 11, divides both $f(x)$ and $g(x)$. Therefore $d(x)=$the *GCD* of $f(x)$ and $g(x)=1$. From equation (ii) we get,

$$11=(x^2+2)-(x+3)(x-3) \quad \ldots \text{(iv)}$$

From (i) and (iv) we have,

$$\begin{aligned}11&=(x^2+2)-(x+3)[(x^3-2x^2+3x-7)-(x-2)(x^2+2)]\\&=(x^2+2)(x+3)(x^3-2x^2+3x-7)+(x^2-x-6)(x^2+2)\\&=(2x^2+x-4)(x^2+2)+(-x-3)x(x^3-2x^2+3x-7)\end{aligned}$$

Therefore,

$$1=\left(\frac{2}{11}x^2+\frac{1}{11}x-\frac{4}{11}\right)(x^2+2)+\left(-\frac{1}{11}x-\frac{3}{11}\right)(x^3-2x^2+3x-7)$$

(b) If $f(x)=x^3-3x^2+2x-6$ and $g(x)=x^3-4x^2+4x-3$, then find the *GCD* of $f(x)$ and $g(x)$.

By division algorithm,

$$x^3-3x^2+2x-6=1(x^3-4x^2+4x-3)+x^2-2x-3$$

$$x^3-4x^2+4x-3=(x-2)(x^2-2x-3)+3x-9$$

$$x^2-2x-3=\left(\frac{1}{3}x+\frac{1}{3}\right)(3x-9)+0$$

Therefore, the *GCD* of $f(x)$ and $g(x)$

$$=\frac{1}{3}(3x-9)=(x-3)$$

$$\begin{aligned}3x-9 &= (x^3-4x^2+4x-3)-(x-2)(x^2-2x-3)\\ &= (x^3-4x^2+4x-3)-(x-2)\times\\ &\qquad [x^3-3x^2+2x-6-(x^3-4x^2+4x-3)]\\ &= (x-1)(x^3-4x^2+4x-3)+(-x+2)(x^3-3x^2+2x-6)\end{aligned}$$

Therefore,

$$(x-3)=\left(\frac{1}{3}x-\frac{1}{3}\right)(x^3-4x^2+4x-3)+\left(-\frac{1}{3}x+\frac{2}{3}\right)(x^3-3x^2+2x-6)$$

5.1.1 The Field *C*

We shall now give a very important application of Theorem 5.6(d) to give the definition and construction of the complex number field ***C***. This is a little digression from our main topic, but it is very much instructive.

Let ***R*** denote the field of all real numbers and ***R***[*x*] the ring of all polynomials over *R* in the indeterminate *x*. This is a commutative ring with unit element (in fact a PID and a UFD).

Let,

$$J=\{(x^2+1)p(x): p(x)\in \mathrm{R}[x]\}$$

It is now easy to see that *J* is an ideal in $R[x]\,(x^2+1)$ is irreducible over ***R.*** Therefore by Theorem 5.6(d) *J* is a maximal ideal in ***R***[*x*]. Now consider the quotient ring $R[x]/J$. Since *J* is a maximal ideal, $R[x]/J$ is a field. This field is called the field of all complex numbers. The elements of $R[x]/J$ are of the form $a + bx + J$ where $a,\ b \in$ ***R***. We write *C* for $R[x]/J$ and $a + bi$ for the element $a + bx + J$.

Let $\bar{R}=\{a+J : a\in R\}$. Then $\bar{R}$ is a field isomorphic to ***R*** and is a subfield of *C*. In fact, if a mapping $\psi: R[x]\to R[x]/J=C$ is defined by,

$$\psi(p(x))=p(x)+J$$

then the restriction of ψ to ***R*** defines an isomorphism of ***R*** onto $\bar{R}$. Now we identity ***R*** with $\bar{R}$ because of the isomorphism and regard ***R*** to be a subfield of *C*, in other words *C* to be a super field (and what is called an extension) of ***R***.

Definition

Let $f(x)=a_0+a_1x+...+a_nx^n$ be a polynomial over a UFD ***R***. The content of the polynomial *f*(*x*) is the *GCD* of $a_0, a_1, ... a_n$. The polynomial *f*(*x*) is said to be *primitive* if the content is 1, in other words, a polynomial $f(x)$ is primitive if the *GCD* of $a_0, a_1, \dots a_n$ is 1.

Theorem 5.7

Let $\boldsymbol{R}$ be a UFD. If $f(x)$ and $g(x)$ are primitive polynomials, then $f(x)g(x)$ is a primitive polynomial.

Proof: Let $f(x) = a_0 + a_1x + \dots + a_nx^n$ and $g(x) = b_0 + b_1x + \dots + b_mx^m$ be primitive polynomials. Suppose to the contrary that $f(x)g(x)$ is not primitive. Then $f(x)g(x)$ is divisible by some prime element p of $\boldsymbol{R}$. Since $f(x)$ is primitive, p does not divide some coefficient a_i of $f(x)$. Let a_i be the first coefficient of $f(x)$ which is not divisible by p. Similarly let b_k be the first coefficient of $g(x)$ which is not divisible by p. Observe that the coefficient of x^{j+k} in $f(x)$ $g(x)$ is given by,

$$c_{j+k} = a_jb_k + (a_{j+1}b_{k-1} + a_{j+2}b_{k-2} + \dots + a_{j+k}b_0) \\ + (a_{j-1}b_{k+1} + a_{j-2}b_{k+2} + \dots + a_0b_{j+k})$$

Now by the choice of b_k, $p \mid b_{k-1}, b_{k-2}, \dots$ so that,

$$p \mid (a_{j+1}b_{k-1} + a_{j+2}b_{k-2} + \dots + a_{j+k}b_0)$$

Similarly by the choice of a_j, $p \mid a_{j-1}, a_{j-2}, \dots$ so that,

$$p \mid (a_{j-1}b_{k+1} + a_{j-2}b_{k+2} + \dots + a_0b_{j+k})$$

By our assumption $p \mid c_{j+k}$. Thus $p \mid a_jb_k$. This contradicts the fact that $p \dagger a_j$ and $p \dagger b_k$ and completes the proof.

We shall complete this section by proving two useful results, namely Gauss Lemma and Eisenstein Criterion. For simplicity we shall assume $\boldsymbol{R}$ the UFD, to be the set $\boldsymbol{Z}$ of all integers and hence its field of quotients will be $\boldsymbol{Q}$, the rationals.

Theorem 5.8 (Gauss Lemma)

If the primitive polynomial $f(x)$ can be factored as the product of two polynomials having rational coefficients then it can be factored as the product of two polynomials having integer coefficients.

Proof: Suppose that $f(x) = u(x)v(x)$ where $u(x)$ and $v(x)$ have rational coefficients. Then we can write $f(x) = (a/b)\lambda(x)\mu(x)$ where a, b are integers and $\lambda(x), \mu(x)$ have integer coefficients and are primitive. The content of each of $f(x)$, $\lambda(x)$ and $\mu(x)$ is 1 since each of them is primitive. Therefore

$a/b = 1$ and hence $f(x) = \lambda(x)\mu(x)$ where $\lambda(x), \mu(x)$ have integer coefficients. This completes the proof.

Theorem 5.9 (The Eisenstein Criterion)

Let $f(x) = a_0 + a_1x + ... + a_nx^n$ be a polynomial with integer coefficients. Suppose that for some prime number p, $p \dagger a_j$, $p \mid a_1$, $p \mid a_2$, ..., $p \mid a_0$, $p^2 \dagger a_0$. Then $f(x)$ is irreducible over the field $\boldsymbol{Q}$.

Proof: Since $p \dagger a_n$, by taking out the *gcd* of the coefficients of $f(x)$ the hypothesis of the theorem is not affected. Therefore without loss of generality we may assume that $f(x)$ is primitive. If $f(x)$ can be factored as a product of two polynomials with rational coefficients, then by Gauss Lemma it follows that it can be factored as the product of two polynomials having integer coefficients. Hence if we assume that $f(x)$ is reducible then,

$$f(x) = (b_0 + b_1x + ... + b_rx^r)(c_0 + c_1x + ... + c_sx^s)$$

where $r > 0$, $s > 0$, $b_0, b_1, ..b_r$, $c_0, c_1, ..., c_s$ are integers. By comparing the coefficients we first get $a_0 = b_0c_0$, since $p \mid a_0$, p must divide at least one of b_0 or c_0. Since $p^2 \dagger a_0$, p cannot divide both b_0 and c_0. Suppose that $p \mid b_0$ and $p \dagger c_0$. Not all the coefficients $b_0, b_1, ...b_r$ can be divisible by p, otherwise all the coefficients of $f(x)$would be divisible by p, which would contradict the fact that $p \dagger a_n$. Let b_k be the first coefficient in,

$$b_0 + b_1x + ... + b_rx^r$$

which is not divisible by p, $k \le r < n$. Thus $p \mid b_{k-1}b_{k-2}, ..., b_0$. But

$$a_k = b_kc_0 + b_{k-1}c_1 + ... + b_0c_k$$

and $p \mid a_k, p \mid b_{k-1}, b_{k-2}, ..., b_0$ so that $p \mid b_kb_0$. This contradicts the fact that $p \dagger c_0$, $p \dagger b_k$.

This contradiction shows that $f(x)$ cannot be factored and hence $f(x)$ is irreducible.

5.2 EXTENSION FIELDS

In the remaining parts of this chapter we shall be concerned with the theory of equations which treats the questions about the roots of polynomials and the algebraic numbers. We shall touch only lightly on the rudiments of this topic and shall not go into the material in its fullest and most general form. For example, we shall not introduce the beautiful theory due to the famous French mathematician Evariste Galois in this introductory text and therefore we shall not be able to supply the proof that the general polynomial of degree $n \geq 5$ is not solvable by radicals.

But even for an elementary study of roots of polynomials we need some basic concepts of field theory. Field theory plays a central role in algebra. The results about them not only find applications in the theory of equations but also in the theory of Numbers and in several other situations.

Therefore before going to define algebraic numbers and roots of polynomials we shall first discuss some concepts regarding extension of fields.

Definition

Let F be a field. A field E is said to be an *extension* of F if E contains F as a subfield.

It should be noted that if E is an extension of a field F then E is a vector space over the field F under addition and scalar multiplication respectively as addition and multiplication in the field.

The *degree* of E over F is defined to be the dimension of E as a vector space over F. Throughout the rest of this chapter we shall denote by F a field, E an extension of F. We write $[E: F]$ to denote the degree of E over F. If $[E: F]$ is finite, i.e. if the dimension of E as a vector space over F is finite, then E is called a finite extension of F and this in most cases is of particular importance to us. We first prove:

Theorem 5.10

If E is a finite extension of F and L is a finite extension of E then L is a finite extension of F and $[L:F] = [L:E][E:F]$

Proof: Let $\alpha_1, \alpha_2, ..., \alpha_n$ be a basis for E over F and $\beta_1, \beta_2, ..., \beta_m$ a basis for L over E. Then we claim that the mn elements $\alpha_i \beta_j$ form a basis of L over F.

Let α be any element of L. Then since $\beta_1, \beta_2, ..., \beta_m$ is a basis for L over E, we can find elements $k_1, k_2, ..., k_m$ in E such that,

$$\alpha = k_1\beta_1 + k_2\beta_2 + ... + k_m\beta_m$$

Again since $\alpha_1, \alpha_2, ..., \alpha_n$ is a basis of E over F, for each k_i we can find elements $f_{i1}, f_{i2}, ..., f_{in}$ in F such that,

$$k_i = f_{i1}\alpha_1 + f_{i2}\alpha_2 + ... + f_{in}\alpha_n$$

So,

$$\begin{aligned}\alpha &= k_1\beta_1 + k_2\beta_2 + \ldots + k_m\beta_m \\ &= \sum_{i=1}^{m}(f_{i1}\alpha_1 + f_{i2}\alpha_2 + \ldots + f_{in}\alpha_n)\beta_i \\ &= \sum_{i=1}^{m}\sum_{j=1}^{n} f_{ij}\alpha_j\beta_i\end{aligned}$$

Further it is left to the reader as an exercise to prove that the mn elements $\alpha_i\beta_j$ are linearly independent. Hence L is a finite extension of F and,

$$[L: F] = mn = [L: E]\ [E: F]$$

We shall be concerned with two questions: Firstly, given an extension E of F and given an element $a \in E$ does there exist a nonzero polynomial $p(x) = a_0 + a_1x + \ldots + a_nx^n \in F[x]$ such that,

$$p(a) = a_0 + a_1a + \ldots + a_na^n = 0\ ?$$

This is not always true. For example, let $E = \boldsymbol{R}$, $F = \boldsymbol{Q}$. Then given $e \in R$ there exists no $p(x) \in \boldsymbol{Q}[x]$ such that $p(e) = 0$. Of course, given $a \in F$ there always exists $p(x) \in F[x]$ such that $p(a) = 0$. For example, take $p(x) = x - a$. Secondly, given a polynomial $p(x) \in F[x]$ does there exist an element $a \in F$ such that $p(a) = 0$?

This is also not always true. For example, let $F = \boldsymbol{Q}$ and $p(x) = x^2 - 2$. As another example, let $F = \boldsymbol{R}$ and $p(x) = x^2 + 1$. Therefore we ask: given $p(x) \in F[x]$ does there exist an extension E of F such that there exists an element $a \in E$ for which $p(a) = 0$? No longer the field F is available to us, in fact it is our prime objective to see whether such a field exists and if so to construct it. Any extension of F does not have this property: For example, let $F = \text{Q}$, $p(x) = x^2 + 1$ and $E = \boldsymbol{R}$. Then although $\boldsymbol{R}$ is an extension of $\boldsymbol{Q}$, there exists no $a \in \mathbf{R}$ such that $a^2 + 1 = 0$.

We shall first dispose off the first question leaving the second question to be discussed later. We first introduce the following definitions.

Definitions

(a) Let $p(x) \in F[x]$. An element a lying in some extension E of F is called a root (or zero) of the polynomial $p(x)$ if $p(a) = 0$. If a is a root of $p(x)$ we sometimes say that a satisfies $p(x)$.

(b) Let E be a given extension of F. An element $a \in E$ is said to be *algebraic* over F if there exists a nonzero polynomial $p(x) \in F[x]$ which a satisfies, i.e. of which a is a root.

(c) An element $a \in E$ is said to be algebraic of degree n over F if it satisfies a nonzero polynomial over F of degree n but no nonzero polynomial of lower degree.

(d) An extension E of F is called an *algebraic extension* of F if every element of E is algebraic over F.

(e) A complex number which is not algebraic is called transcendental.

(f) An algebraic number a is called an algebraic integer if it satisfied a polynomial of the form $x^n + a_1x^{n-1} + ... + a_n$ where $a_1, a_2, ..., a_n$ are integers.

We now state a very important result known as *Gelfand-Schneider* Theorem. We cannot prove this theorem at this stage and will not try to prove the theorem in this introductory book. The theorem states as follows:

Theorem 5.11

Gelfand-Schneider Theorem: Let α and β be algebraic numbers and let β be irrational. Then α^{β} transcendental.

Note that every rational number *a/b* is algebraic, it is the root of the polynomial *bx–a*. Similarly complex number i being the root of $x^2 + 1$ is algebraic. But while defining algebraic and transcendental numbers it is pertinent question to ask: Whether transcendental numbers exist.

One and the best way of showing this is to demonstrate that some specific number is transcendental.

In 1851 *Liouville* gave a characterisation of a complex number to be algebraic. Using this he was able to write down a large collection of transcendental numbers and this settled the question of their existence. The next problem, which was of great importance, was whether some specific and familiar number was transcendental. The first success was due to *Hermite* in *1873* who proved that e is transcendental and his proof was simplified by Hilbert afterwards. The next success in this direction is probably due to *Lindemann* who in 1882 proved that π is transcendental. In 1934, working independently, *Gelfand* and *Schneider* proved the *Gelfand-Schneider* theorem which answered in the affirmative the question raised by Hilbert whether $2^{\sqrt{2}}$ was transcendental.

We now turn to the second question and try to find an extension E of the field F in which a given polynomial $p(x) \in F[x]$ has a root. We shall also try to find the extension in which the polynomial will have all the roots.

We begin with the important result known as the Remainder theorem which will be needed in the sequel.

Theorem 5.12 (Remainder Theorem)

Let $p(x) \in F[x]$. Let E be an extension of F and let $b \in E$. Then $p(x) = (x-b)q(x) + p(b)$ where $q(x) \in E[x]$ and deg $q(x)$ = deg $p(x) - 1$.

Proof: Since $F \subset E$, $F[x] \subset E[x]$ and hence we can consider $p(x)$ to lie in $E[x]$. By applying division algorithm for polynomials in $E[x]$ we get,

$$p(x) = (x-b)q(x) + r$$

where $q(x) \in E[x]$ and $r = 0$ or deg $r <$ deg $(x-b) = 1$. Thus either $r = 0$ or deg $r = 0$, so in either case r is an element of E. Now $p(b) = (b-b)q(x) + r = 0 + r = r$.

Therefore,

$$p(x) = (x-b)q(x) + r$$

where $q(x) \in E[x]$ and deg $q(x)$ is one less than deg $p(x)$ and this completes the proof.

Corollary

Let E be an extension of F and $p(x) \in F[x]$. $a \in E$ is a root of $p(x)$ if and only if $(x-a) \mid p(x)$.

Proof: By Remainder theorem we have,

$$p(x) = (x-a)q(x) + p(a)$$

If $a \in E$ is a root of $p(x)$ then $p(a) = 0$. Hence, $p(x) = (x-a)q(x)$ and thus $(x-a) \mid p(x)$. Conversely, if $(x-a) \mid p(x)$ then $p(a) = 0$ and hence a is a root of $p(x)$.

Definition

An element $a \in E$ is said to be a root of $p(x) \in F[x]$ of multiplicity m if $(x-a)^m \mid p(x)$ and $(x-a)^{m+1} \nmid p(x)$.

If $a \in E$ is a root of $p(x) \in F[x]$ of multiplicity m, then it is counted as m roots. Roots of multiplicity 1 are the usual roots and these are sometimes called simple roots, the roots of multiplicity 2 are called double roots.

Now we ask two questions: How many roots can a polynomial have in a given field, and whether a given polynomial has a root at all. We shall answer these questions in the following three theorems.

Theorem 5.13

A polynomial of degree n over a field can have at most n roots in any extension field.

Proof: Let $p(x)$ be a polynomial of degree n over a field F. We shall prove the theorem by induction on n. If $n = 1$ then $p(x)$ is of the form $\alpha x+\beta$ where $\alpha,\beta \in F$ and $\alpha \neq 0$. If $a=-\beta/\alpha \in F$ then $p(x)$ has the unique root $a \in E$ and the theorem holds for $n = 1$.

We now make the induction hypothesis that the result holds for all polynomials of degree less than n over any field. Now assume that $p(x)$ is of degree n over F and that E is an extension of F. If $p(x)$ has no roots in E then we are through because the number of roots in E being zero is obviously less than n. So we suppose that $p(x)$ has at least one root $a \in E$ and it is a root of multiplicity m. Since $(x-a)^m \mid p(x)$ it follows that $m \leq n$. Now $p(x)=(x-a)^m q(x)$ where $q(x) \in E(x)$ and is of degree n–m. Since $(x-a)^{m+1} \dagger p(x)$ we have $(x-a) \dagger q(x)$ and hence a is not a root of $q(x)$. If $b \neq a$ is a root in E of $p(x)$ then $0=p(b)=(b-a)^m q(b)$, however since $b-a \neq 0$ and since E is a field we have $q(b)=0$. This means that any root of $p(x)$ in E other than a is a root of $q(x)$. Now deg $q(x) = n-m < n$. Therefore by our induction hypothesis, $q(x)$ has at most $(n-m)$ roots in E, this together with the other root a of $p(x)$ counted m times says that $p(x)$ has at most $m + (n-m) = n$ roots in E. This completes the proof of the theorem.

Theorem 5.14

If $p(x)$ is a polynomial in $F[x]$ of degree $n \geq 1$ and is irreducible over F, then there exists an extension E of F such that $[E: F] = n$ in which $p(x)$ has a root.

Proof: Let $V = (p(x))$ be the ideal of $F[x]$ generated by $p(x)$. Then observe that V is a maximal ideal of $F[x]$ and consequently $E = F[x]/V$ is a field. We claim that this field E is our required extension. To establish our claim we have to, first of all, show that E is an extension of F. Strictly speaking E is not really an extension of F, it is an extension of an isomorphic image of F. To show this let,

$$\overline{F}=\{a+V : a \in F\}$$

Let $\psi : F[x] \to E = F[x]/V$ be defined by

$$\psi(f(x)) = f(x) + V$$

Then the restriction of ψ induces an isomorphism of F onto F and E is an extension of F.

Now E is a finite extension of F of degree $n = \deg p(x)$, because the elements $1 + V, x + V, (x + V)^2 = x^2 + V, \ldots, (x + V)^I = x^i + V, \ldots, (x + V)^{n-1} = x^{n-1} + V$ form a basis of E over F. Let us put $\psi(x) = x + V = a \in E$. Observe that given $f(x) \in F[x]$, $\psi(f(x)) = f(a)$.

For since ψ is a homomorphism, if $f(x) = \beta_0 + \beta_1 x + \ldots + \beta_k x^k$ then $\psi(f(x)) = \psi(\beta_0) + \psi(\beta_1)\psi(x) + \ldots + \psi(\beta_k)\psi(x)^k$ and identifying $\psi(\beta)$ with β we get,

$$\psi(f(x)) = \beta_0 + \beta_1 a + \ldots + \beta_k a^k = f(a)$$

In particular, since $p(x) \in V$, $\psi(p(x)) = 0$. But $\psi(p(x)) = p(a)$. Therefore $p(a) = 0$ and thus the element $a = \psi(x) \in E$ is a root of $p(x) \in F[x]$ and this completes the proof of the theorem.

Corollary

If $f(x) \in F[x]$ then there is a finite extension E of F in which $f(x)$ has a root and $[E: F]$ $[E : F] \leq \deg f(x)$.

Proof: Let $p(x)$ be an irreducible factor of $f(x)$. Observe that any root of $p(x)$ is a root of $f(x)$. By Theorem (5.14) there is an extension E of F with $[E: F] = \deg p(x) \leq \deg f(x)$ in which $p(x)$ and consequently $f(x)$ has a root.

Theorem 5.15

Let $f(x) \in F[x]$ be of degree $n \geq 1$. Then there is an extension E of F of degree at most $n!$ in which $f(x)$ has n roots (and so, all the roots).

Proof: By the above corollary there is an extension E_0 of F with $[E_0 : F] \leq n$ in which $f(x)$ has a root α. Therefore in $E_0[x]$, $f(x)$ factors as $f(x) = (x - \alpha)q(x)$, where $\deg q(x) = n - 1$. Continuing in the above manner, there is an extension E of E_0 of degree at most $(n-1)$ in which $q(x)$ has $(n-1)$ roots. Since any root of $f(x)$ is either a or a root of $q(x)$, in E we obtain all n roots of $f(x)$. Now,

$$[E : F] = [E : E_0][E_0 : F] = (n-1)!n = n!$$

and this completes the proof.

Remark

Thus it follows from the above theorem that given $f(x) \in F[x]$ and deg $f(x)$ = n, there exists a finite extension E of F in which $f(x)$ has all the roots. This means that $f(x)$ splits up completely over E as a product of linear factors. Since a finite extension of F exists with this property, a finite extension of F exists with minimal degree which has this property. Such a field is called a *splitting field* over F of $f(x)$. In other words, if $f(x) \in F[x]$, a finite extension E of F is said to be splitting field over F for $f(x)$ if over E but not over any proper subfield of E, $f(x)$ can be factored as a product of linear factors.

5.3 ROOTS OF POLYNOMIALS

In this section, we shall be concerned with the polynomials over the complex field C and in particular, over the real field $\boldsymbol{R}$, rational field $\boldsymbol{Q}$ and ring of integers $\boldsymbol{Z}$. We shall discuss some important properties of roots of such polynomials and relation between the roots and the coefficients.

There are several methods or rules given by several eminent mathematicians such as: Gauss, Newton, Euler, Fourier, Ferrari, Budan, Strum, Horner for determination of roots of general quadratic, cubic and biquadratic equations.

We shall not discuss any of these methods in this book. The interested reader may refer to any good text on Theory of Equations. It should be noted that the solutions of polynomial equations of degrees higher than 4 are given only in some special cases. For example, the solutions of the equation $x^{17} - 1 = 0$ was discovered by the famous mathematician Guass. The reason is that any general polynomial of degree greater than or equal to 5 is not solvable by radicals. We cannot prove this result in this introductory book because it needs the ideas of Galois theory which we do not intend to discuss here. The reader is referred to any advanced text on Abstract algebra which deals with Galois theory.

Another important result which is called the fundamental theorem of algebra states that the complex number field $\boldsymbol{C}$ is algebraically closed, in other words every polynomial of degree n over C (or in particular, over $\boldsymbol{R}$ or $\boldsymbol{Q}$ or $\boldsymbol{Z}$) has exactly n roots (and hence all the roots) in $\boldsymbol{C}$. The proof of this result uses techniques of complex Analysis and we do not intend to prove this result also here.

The following theorems (Theorems 5.16–5.20) describe some basic properties of roots of polynomials and the relation between roots and coefficients.

Theorem 5.16

Let $f(x)=a_0+a_1x+...+a_nx^n$ be a polynomial over **Z**. If $\frac{p}{q}\in Q$ is a root of $f(x)$ then $p\mid a_0$ and $q\mid a_n$.

Proof: If $\frac{p}{q}$ is a root of $f(x)$ then,

$$a_0+a_1\left(\frac{p}{q}\right)+...+a_n\left(\frac{p}{q}\right)^n=0$$

Or $a_0q^n+a_1pq^{n-1}+...+a_np^n=0$

Or $a_0q^n=-a_1pq^{n-1}-...-a_np^n$

$$=p(-a_1q^{n-1}-...-a_{n-1}p^{n-2}q-a_np^{n-1})$$

Thus $p\mid a_0q^n$ and since p and q are relatively prime, $p\mid a_0$. Also we have,

$$\begin{aligned}a_np^n&=-a_0q^n-a_1pq^{n-1}-...-a_{n-1}p^{n-1}q\\&=q(-a_0q^{n-1}-a_1pq^{n-2}-...-a_{n-1}p^{n-1})\end{aligned}$$

Thus $q\mid a_np^n$, since p and q are relatively prime, $q\mid a_n$ and this completes the proof.

Theorem 5.17

If a polynomial $f(x)$ with real coefficients has $\alpha+i\beta$ as one of the roots then $\alpha-i\beta$ is also a root, in other words if complex roots occur then they occur in pair.

Proof: Let $f(x)\in R[x]$ be of degree n and let $\alpha+i\beta$ be a root of $f(x)$, i.e. $f(\alpha+i\beta)=0$. Observe that,

$$\{x-(\alpha+i\beta)\}\{x-(\alpha-i\beta)\}=(x-\alpha)^2+\beta^2$$

Now divide by $f(x)$ by $\{(x-\alpha)^2+\beta^2\}$. By division algorithm we obtain,

$$f(x)=\{(x-\alpha)^2+\beta^2\}\phi(x)+(k_1x+k_2)$$

where $\phi(x)$ is a polynomial over **R** of degree (n–2) and k_1, k_2 are two real constants. Now,

$\alpha+i\beta$ is a root of $f(x)$, i.e. $f(\alpha+i\beta)=0$.

Therefore,

$$\begin{aligned}0 &= \{(\alpha+i\beta-\alpha)^2+\beta^2\}\phi(\alpha+i\beta)+k_1(\alpha+i\beta)+k_2\\ &= (-\beta^2+\beta^2)\phi(\alpha+i\beta)+k_1(\alpha+i\beta)+k_2\\ &= k_1(\alpha+i\beta)+k_2\\ &= (k_1\alpha+k_2)+i\beta k_1\end{aligned}$$

Consequently $(k_1\alpha+k_2)=0$
and $k_1\beta=0$ where $\beta\neq 0$.

Therefore, $\beta\neq 0.$ and so $k_2=0$.

Hence,

$$\begin{aligned}f(x) &= \{(x-\alpha)^2+\beta^2\}\phi(x)\\ &= \{x-(\alpha+i\beta)\}\{x-(\alpha-i\beta)\}\phi(x)\end{aligned}$$

Now since $\{x-(\alpha-i\beta)\}$ is also a factor of $f(x)$ this shows that $(\alpha-i\beta)$ is also a root of $f(x)$.

This completes the proof.

Theorem 5.18

Let $f(x)$ be a polynomial with rational coefficients. Then its irrational roots occur in pair.

Proof: Let $f(x)$ be a polynomial of degree n over the field $\boldsymbol{Q}$ of rational numbers. Let $a+\sqrt{b}$ be a irrational root of $f(x)$. Observe that,

$$(x-a-\sqrt{b})(x-a+\sqrt{b})=(x-a)^2-b$$

Divide $f(x)$ by $(x-a)^2-b$. Let $f(x)=\{(x-a)^2-b\}\phi(x)+k_1x+k_2$ where k_1 and k_2 are rationals. Since $a+\sqrt{b}$ is a root of $f(x)$ we have,

$$\begin{aligned}0=f(a+\sqrt{b}) &= \{(a+\sqrt{b}-a)^2-b\}\phi(a+\sqrt{b})+k_1(a+\sqrt{b})+k_2\\ &= k_1(a+\sqrt{b})+k_2\\ &= (k_1a+k_2)+k_1\sqrt{b}\end{aligned}$$

Therefore, $k_1a+k_2=0$ and $k_1=0$.

Consequently $k_2=0$.

Thus, $f(x)=\{(x-a)^2-b\}\phi(x)$

$$=[\{x-(a+\sqrt{b})\}\{x-(a-\sqrt{b})\}]\phi(x)$$

Therefore $a-\sqrt{b}$ is also root of $f(x)$ and this completes the proof.

In the following theorem we shall discuss some properties of the polynomial $x^n - 1$.

Theorem 5.19

(i) If α is a root of $x^n - 1$ then α^r is also a root of $x^n - 1$ where r is a positive integer.

(ii) If m and n are relatively prime then the polynomials $x^n - 1$ and $x^m - 1$ do not have any common root except 1.

(iii) If α is an imaginary root of $x^n - 1$ where n is prime then all its roots are $1, \alpha, \alpha^2, ..., \alpha^{n-1}$.

Proof:

(i) Since α is a root of $x^n - 1$ we have $\alpha^n = 1$. Therefore

$$(\alpha^r)^n = (\alpha^n)^r = 1^r = 1$$

and hence α^r is a root of $x^n - 1$.

(ii) Since m and n are relatively prime we can choose integers a and$b$ such that $mb - na = \pm 1$. Assume that α is a root of both the polynomials $x^n - 1$ and $x^m - 1$. Then

$\alpha^m = 1$ and $\alpha^n = 1$.

By (i) we have $\alpha^{mb} = 1$ and $\alpha^{mb} = 1$. Therefore $\alpha^{mb-na} = 1$, i.e. $\alpha = \pm 1$, i.e., $\alpha = 1$ and this completes the proof.

(iii) Let α be a root of $x^n - 1$. By (i) α^r is also a root of $x^n - 1$where r is a positive integer.

Since $x^n - 1$ is a polynomial of degree n, it has exactly n roots in the field of complex numbers. Again $1, \alpha, \alpha^2, ...$ are the roots of $x^n - 1$. In order to prove the theorem it is enough to show that $1, \alpha, \alpha^2, ..., \alpha^{n-1}$ are all distinct. If possible assume that $\alpha^r = \alpha^m$ with $r<m<n$. Then $\alpha^{m-r} = 1$. Hence α is a root of $x^{m-r} - 1$. In other words α is a common root of $x^n - 1$ and $x^{m-r} - 1$. But m and r are both less than n and n is a prime, $(m - r)$ and n are relatively prime. Therefore there is no root common to $x^n - 1$ and $x^{m-r} - 1$ except 1. Hence $\alpha^r \neq \alpha^m$ for $r \neq m$ and this completes the proof of the theorem.

The following theorem gives the relations between the roots and coefficients of a polynomial.

Theorem 5.20

Let $f(x)=a_0+a_1x+...+a_nx^n$ be a polynomial over the field ***Q*** or ***R*** or ***C***. Let $\alpha_1,\alpha_2,...,\alpha_n$ be all the distinct roots of $f(x)$ all lying in C. Let,

$$S_1=\alpha_1+\alpha_2+...+\alpha_n=\sum\alpha_1$$

$$S_2=\alpha_1\alpha_2+...=\sum\alpha_1\alpha_2$$

$$S_3=\sum\alpha_1\alpha_2\alpha_3$$

....

....

$$S_n=\alpha_1\alpha_2...\alpha_n$$

Then,

$$S_j=(-1)^j\frac{a_{n-j}}{a_n},\quad j=1,2,...,n$$

Proof: Since $\alpha_1,\alpha_2,...,\alpha_n$ are the roots of $f(x)$ we have

$$f(x)=a^n(x-\alpha_1)(x-\alpha_2)...(x-\alpha_n)$$

Or
$$\begin{aligned}&a_0+a_1x+...+a_nx^n\\&=a_n(x-\alpha_1)(x-\alpha_2)...(x-\alpha_n)\\&=a_n[x^n-(\sum\alpha_i)x^{n-1}+(\sum\alpha_1\alpha_2)x^{n-2}+...+(-1)^n\alpha_1\alpha_2...\alpha_n)]\\&=[\alpha^n-S_1x^{n-1}+(-1)^2S_2x^{n-2}+...+(-1)^nS_n]\end{aligned}$$

Now comparing the coefficients of various powers of x on both sides we get,

$$a_{n-1}=-a_nS_1 \text{ or } S_1=-\frac{a_{n-1}}{a_n}$$

$$a_{n-2}=(-1)^2a_nS_2 \text{ or } S_2=(-1)^2\frac{a_{n-2}}{a_n}$$

$$a_{n-j}=(-1)^ja_nS_j \text{ or } S_j=(-1)^j\frac{a_{n-j}}{a_n}$$

5.4 SOLUTION OF CUBIC AND BI QUADRATIC EQUATION

We shall discuss below the solutions of general cubic equations by radical method. The general cubic equation is given by,

$$x^3+px^2+Qx+R=0$$

This can be transformed into a simpler equation in the form,

$$x^3 + px + r = 0$$

In order to solve this equation, put $x = y + z$. Then we have,

$$\begin{aligned} x^3 &= y^3 + z^3 + 3yz(y+z) \\ &= y^3 + z^3 + 3yzx \end{aligned}$$

and the equation takes the form,

$$y^3 + z^3 + (3yz + q)x + r = 0.$$

At present, y and z are any two quantities subject to the condition that their sum is equal to one of the roots of the given equation. If we further assume that they satisfy the equation $3yz + q = 0$, then they are completely determined. We thus obtain,

$$y^3 + z^3 = -r,$$

$$y^3 z^3 = -\frac{q^3}{27}.$$

Hence, y^3 and z^3 are the roots of the quadratic equation,

$$t^2 + rt - \frac{q^3}{27} = 0.$$

Solving this equation we get,

$$y^3 = -\frac{r}{2} + \sqrt{\frac{r^2}{4} + \frac{q^3}{27}}$$

$$z^3 = -\frac{r}{2} - \sqrt{\frac{r^2}{4} + \frac{q^3}{27}}.$$

Thus we get the value of x from the relation $x = y + z$ as follows:

$$x = \left\{-\frac{r}{2} + \sqrt{\frac{r^2}{4} + \frac{q^3}{27}}\right\}^{1/3} + \left\{-\frac{r}{2} - \sqrt{\frac{-r}{4} + \frac{q^3}{27}}\right\}$$

We shall next discuss the solution of biquadratic equation by radical method. This was first obtained by Ferrari.

Let the equation be,

$$x^4 + 2px^3 + qx^2 + 2rx + s = 0$$

Add $(ax+b)^2$ to both sides, the numbers a and b to be determined later so that the left hand side becomes a perfect square.

Then we have,

$$x^4+2px^2+(q+a^2)x^2+2(r+ab)x+s+b^2=(ax+b)^2$$

Suppose that the left hand side is equal to $(x^2+px+k)^2$. Then by comparing the coefficients on both sides we obtain,

$$p^2+2k=q+a^2,\quad pk=r+ab,\quad k^2=s+b^2$$

By eliminating a and b from the equations, we obtain,

$$(pk-r)^2=(2k+p^2-q)(k^2-s),$$

Or $2k^2-qk^2+2(pr-s)k-p^2s+qs-r^2$

For, this cubic equation, one real value of k can always be found, thus a and b are known.

Also

$$(x^2+px+k)^2=(ax+b)^2$$

Or $x^2+px+k=\pm(ax+b)$

Now the values of x are to be obtained from the two quadratic equations,

$$x^2+(p-a)x+(k-b)=0$$

and $x^2+(p+a)x+(k+b)=0$

and this completes the problem.

Example 5.4

(a) Determine a polynomial whose roots are –3, –1, 4, 5. Since – 3, –1, 4 and 5 are the roots, the polynomial $f(x)$ is given by,

$$\begin{aligned} f(x) &= (x+3)(x+1)(x-4)(x-5) \\ &= x^4-5x^3-13x^2+53x+60 \end{aligned}$$

(b) If one root of $x^4+2x^3-5x^2+6x+2$ is $-2+\sqrt{3}$ find the other roots.

Since irrational roots occur in pair and since $-2+\sqrt{3}$ is a root, $-2-\sqrt{3}$ is another root. Therefore, $(x+2+\sqrt{3})(x+2-\sqrt{3})=x^2+4x+1$ is a factor of the given polynomial. Dividing the given polynomial by x^2+4x+1 we get x^2-2x+2. Hence the other roots are the roots of x^2-2x+2, given by,

$$\frac{2 \pm \sqrt{4-8}}{2} = 1 \pm i$$

Observe that the two complex roots also occur in pair.

(c) If sum of two roots of $x^3 - 5x^2 - 16x + 80$ is zero, then find all its roots.

This is a cubic equation with integer coefficients and hence it has exactly three roots in the complex field. Let the roots be a, b and c. Then

$$a + b + c = 5 \qquad \dots (1)$$

$$ab + bc + ca = -16 \qquad \dots (2)$$

$$abc = -80 \qquad \dots (3)$$

According to the given condition we may write

$$b + c = 0 \qquad \dots (4)$$

(d) It now follows from (1) and (4) that $a = 5$. Then from (3) we get $bc = -16$. Therefore,

$b - c = \sqrt{(b+c)^2 - 4bc} = \pm 8$ and hence $b = 4$ and $c = -4$.. So all the roots are 5, 4, –4.

If the roots of the polynomial $27x^3 + 42x^2 - 28x - 8$ are in GP then find all the roots.

Since the roots are in GP we can assume without any loss of generality that they are $\frac{a}{b}, a$, and ab

Therefore,

$$\frac{a}{b} + a + ab = \frac{42}{27} \qquad \dots (1)$$

$$\frac{a^2}{b} + a^2 + a^2 b = -\frac{28}{27} \qquad \dots (2)$$

$$\frac{a}{b}.a.ab = \frac{8}{27} \qquad \dots (3)$$

$\therefore a = \frac{2}{3}$ putting this value in (1)

$$\frac{1}{b} + 1 + b = -\frac{21}{9}$$

Or $b^2 + \frac{10}{3} b + 1 = 0$

Or $b = -3$ or $-\frac{1}{3}$

Hence the roots are $-2, \frac{2}{3}, -\frac{2}{9}$.

(e) If a, b, c are the roots of $x^3 + px^2 + qx + r$ then find the value $\sum a^2b$.

Now $a+b+c=-p$

$ab+bc+ca=q$

$abc=-r$

Since $(a+b+c)(ab+bc+ca)=\sum a^2b+3abc$ we have,

$$\sum a^2b=\sum a\sum ab-3abc$$
$$=-p.q-3.(-r)$$
$$=-pq+3r$$

(f) If $\alpha_1, \alpha_2, ..., \alpha_n$ are the roots of:

$x^n + p_1x^{n-1} + ... + p_n$ then find the values of:

$\sum \frac{1}{\alpha_1}$ and $\sum \frac{1}{\alpha_1\alpha_2}$

We know that,

$$\alpha_3\alpha_4...\alpha_n + \alpha_1\alpha_4...\alpha_n + ... = (-1)^{n-2} p_{n-2} \quad ... (1)$$

$$\alpha_2\alpha_3...\alpha_n + \alpha_1\alpha_3...\alpha_n + ... = (-1)^{n-1} p_{n-1} \quad ... (2)$$

$$\alpha_1\alpha_2.\alpha_3...\alpha_n = (-1)^n p_n \quad ... (3)$$

By dividing (2) by (3) we get,

$$\sum \frac{1}{\alpha_1} = \frac{1}{\alpha_1} + \frac{1}{\alpha_2} + ... + \frac{1}{\alpha_n} = \frac{p_{n-1}}{p_n}.$$

(g) If a, b, c are the roots of $x^3 + px^2 + qx + r$, then find the value of $\sum \frac{b^2+c^2}{bc}$

Since a, b, c are the roots of $x^3 + px^2 + qx + r$, we have

$\sum a=-p$

$\sum ab=q$

$abc=-r$

Now,

$$\sum \frac{b^2+c^2}{bc} = \frac{\sum a^2 b}{abc}$$

Again $\sum a \sum ab = \sum a^2 b + 3abc.$

Therefore,

$$\sum \frac{b^2+c^2}{bc} = \frac{\sum a \sum ab - 3abc}{abc}$$

$$= \frac{-p.q+3r}{-r}$$

$$= \frac{pq}{r} - 3$$

(h) If the roots of the polynomial,

$x^n + p_1 x^{n-1} + ... + p_n$ are in AP then find them.

Let the roots of the given polynomial be,

$a, a+d, a+2d, ..., a+(n-1)d$

Then,

$$a+(a+d)+...+\{a+(n-1)d\}$$

$$= na + \frac{n(n-1)}{2} d$$

Therefore,

$$na + \frac{n(n-1)}{2} d = -p$$

Now,

$$a^2 + (a+d)^2 + ... + \{a+(n-1)d\}^2$$

$$= [a+(a+d)+...+\{a+(n-1)d\}]^2 - 2\sum a(a+d)$$

$$= (-p_1)^2 - 2p_2 = p_1^2 - 2p_2 \qquad \text{... (2)}$$

Now multiply (2) by n and from this subtract the square of (1). This gives,

$$d^2 \left\{ \frac{n^2(n-1)(2n-1)}{6} - \frac{n^2(n-1)^?}{4} \right\} = (p_1^2 - 2p_2)n - p_1^2$$

$$\text{Or } d^2 = \frac{12\{(n-1)p_1^2 - 2np_2\}}{n^2(n^2-1)}$$

This gives d, putting this value of d in (1) we get a, and consequently we get all the roots.

(i) If sum of two roots of $x^3 + px^2 + qx + r$ is equal to third one prove that $p^3 - 4pq + 8r = 0$.

We have $\alpha_1 + \alpha_2 = \alpha_3$

$\alpha_1 + \alpha_2 + \alpha_3 = -p$ (1)

$\alpha_1\alpha_2 + \alpha_2\alpha_3 + \alpha_3\alpha_1 = q$ (2)

$\alpha_1\alpha_2\alpha_3 = -r$ (3)

Since $\alpha_1 + \alpha_2 = \alpha_3$ from (1), we get

$$\alpha_3 = -\frac{p}{2}$$

From (2) it follows that,

$\alpha_1\alpha_2 + (\alpha_1 + \alpha_2)\alpha_3 = q$

Or $\alpha_1\alpha_2 + \alpha_3^2 = q$

Or $\alpha_1\alpha_2 + \dfrac{p^2}{4} = q$

Or $\alpha_1\alpha_2 = q - \dfrac{p^2}{4}$

From (3) we get

$$\left(q - \frac{p^2}{4}\right)\left(-\frac{p}{2}\right) = -r$$

Or $(4q - p^2)(-p) = -8r$

Or $p^3 - 4pq + 8r = 0$

(j) If the sum of the roots of $2x^3 - 3x^2 + kx - 1$ is 1, then find the value of k and also find all the roots of the polynomial.

Let $\alpha_1, \alpha_2, \alpha_3$ be the roots of $2x^3 - 3x^2 + kx - 1$.

Now

$\alpha_1 + \alpha_2 + \alpha_3 = \dfrac{3}{2}$ … (1)

$\alpha_1\alpha_2 + \alpha_2\alpha_3 + \alpha_3\alpha_1 = \dfrac{k}{2}$ … (2)

$\alpha_1\alpha_2\alpha_3 = \dfrac{1}{2}$ … (3)

$\alpha_1 + \alpha_2 = 1$ ... (4)

From (1) and (4) we get,

$\alpha_3 = \frac{3}{2} - 1 = \frac{1}{2}$

From (3) $\alpha_1\alpha_2 = 1$ ... (5)

From (4) and (5) we get

$(\alpha_1 - \alpha_2)^2 = (\alpha_1 + \alpha_2)^2 - 4\alpha_1\alpha_2 = 1 - 4 = -3$

Thus $\alpha_1 - \alpha_2 = \pm\sqrt{3}i.$

Therefore $\alpha_1, \alpha_2, \alpha_3$ are respectively,

$\frac{1}{2}, \frac{1}{2}(1+\sqrt{3}i), \frac{1}{2}(1-\sqrt{3}i)$

And

$\alpha_1\alpha_2 + \alpha_3(\alpha_1 + \alpha_2) = \frac{k}{2}$

Or $1 + \frac{1}{2} = \frac{k}{2}$ or $k = 3.$

(k) Find the roots of:

$2x^4 - 11x^3 + 17x^2 - 11x + 15$ in the rational field and in the complex field.

If $(p,q) = 1$ and q is positive and if $p/q \in Q$ is a root of the given polynomial then $p\,|\,15$ and $q\,|\,2$. Hence,

$p = \pm 1,\ \pm 3,\ \pm 5,\ \pm 15,\ q = 1,\ 2$

So the possible roots are:

$\pm 1, \pm\frac{1}{2}, \pm 3, \pm\frac{3}{2}, \pm 5, \pm\frac{5}{2}, \pm 15, \pm\frac{15}{2}$

But among these by verification one can observe that only $f(3) = 0$ and $f(5/2) = 0$. Hence all the rational roots are 3 and 5/2.

Now

$f(x) = (x-3)(2x-5)(x^2+1)$

Hence the roots in the complex field which are not rationals are $\pm i$. Observe that this is a polynomial over the commutative ring Z, it has two roots in Q and all four roots in C. Also observe that complex roots occur in pairs.

(l) Find the rational roots $6x^4 - 7x^3 + 6x^2 - 1$. If $\frac{p}{q} \in Q$ is a root of the given polynomial then $p \mid -1$ and $q \mid 6$. Therefore,
$p = \pm 1$ and $q = 1, 2, 3, 6$.
Hence the possible rational roots are,
$\pm 1, \pm \frac{1}{2}, \pm \frac{1}{3}, \pm \frac{1}{6}$
But it can be checked that only $\frac{1}{2}$ and $\frac{1}{3}$ are the rational roots of the given polynomial and
$f(x) = 6\left(x - \frac{1}{2}\right)\left(x - \frac{1}{3}\right)(x^2 - x - 1)$
The other two roots are real but irrational and these are:
$\frac{1 \pm \sqrt{2}}{2}$

(m) Find the roots of the cubic equation
$x^3 - 15x - 126 = 0$
Let $x = p^{1/3} + q^{1/3}$
Then $x^3 = p + q + 3p^{1/3}q^{1/3}(p^{1/3} + q^{1/3})$
Or $x^3 - 3p^{1/3}q^{1/3}x - (p + q) = 0$
By comparing this with the given equation we get,
$p + q = 126$
$p^{1/3}q^{1/3} = 5$
Or $pq = 125$
Therefore p and q are the roots of:
$t^2 - 126t + 125 = 0$
Now, $0 = t^2 - 126t + 125 = t^2 - t - 125t + 125 = (t-1)(t-125)$
Hence, $p = 125,\ q = 1$.
So, $p^{1/3} = 5,\ 5w,\ 5w^2$
$q^{1/3} = 1,\ w^2,\ w$ (since $p^{1/3}q^{1/3} = 5$)
Therefore, the roots of the given equation are:
$5 + 1,\ 5w + 5w^2,\ 5w^2 + w$
Or $6,\ -3 + i2\sqrt{3},\ -3 - i2\sqrt{3}$

(n) Find the roots of: $x^3-3x+1=0$

If $x=p^{1/3}+q^{1/3}$ then,

$x^3-3p^{1/3}q^{1/3}x-(p+q)=0.$

By comparing the above equation with the given equation we get,

$p+q=1$

$pq=1$

Therefore, p and q are the roots of the quadratic equation $t^2+t+1=0$.

Therefore,

$$t=\frac{-1+i\sqrt{3}}{2}=\cos\frac{2\pi}{3}\pm i\sin\frac{2\pi}{3}$$

$$\text{Hence, } p^{1/3}=\cos\frac{\frac{2\pi}{3}+2\pi r}{3}+i\sin\frac{\frac{2\pi}{3}+2\pi r}{3}$$

$$q^{1/3}=\cos\frac{\frac{2\pi}{3}+2\pi r}{3}-i\sin\frac{\frac{2\pi}{3}+2\pi r}{3}$$

Therefore, the roots of the given equation are:

$$2\cos\frac{\frac{2\pi}{3}+2\pi r}{3},\quad r=0,1,2$$

Or $2\cos\frac{2\pi}{9},\ 2\cos\frac{8\pi}{9},\ 2\cos\frac{14\pi}{9}$.

EXERCISES

1. Determine the polynomials whose roots are:
   (i) $1+\sqrt{2},3,-1$
   (ii) $1,\ -2,\ i$
   (iii) $1+5i,\ 5,\ -i$
   (iv) $-3,\ -1,\ \frac{5}{3}$
2. Find the GCD of the following polynomial over the field of rational numbers:
   (i) $x^3-6x^2+x+4,\ x^5-6x+1$
   (ii) $x^2+1,\ x^4+x^3+x+1$
   (iii) $x^4+x^3+2x^2+x+1,\ x^3-1$
3. If the sum of zeros of $x^3-5x^2-16x+80$ is zero then determine all its zeros.

4. If the zeros of the polynomial $x^3 - px^2 + qx - r$ are in H.P., then prove that:

   $27r^2 - 9qr + 2q^3 = 0$

5. If the zeros of the polynomial:

   $x^3 + 3px^2 + 3qx + r$ are in G.P., then prove that $p^3r - q^3 = 0$

6. In a field F if the zeros of the polynomial $x^3 + qx + r$ are α, β, γ, then prove that the zeros of the polynomial $x^3 + 6qx^2 + 9q^2x + 4q^3 + 27r^3$ are:

   $(\alpha - \beta)^2,\ (\beta - \gamma)^2,\ (\gamma - \alpha)^2$

7. Divide $3x^4 - 5x^3 + 10x^2 + 11x - 61$ by $(x - 3)$

8. By synthetic division divide:

   $x^5 - 4x^4 + 7x^3 + 11x - 13$ by $(x - 5)$

9. If one root of $x^3 + 4x^2 + x - 6 = 0$ is –2, by synthetic division find the other two roots.

10. If one root of $x^3 - 2x^2 - 2x - 3 = 0$ is 3 then by synthetic division find the other two roots.

11. Prove that:

    (i) $x^2 + x + 1$ is irreducible over Z_2

    (ii) $x^2 + 1$ is irreducible over Z_7

    (iii) $x^3 - 9$ is irreducible over Z_{31}

    (iv) $8x^3 - 6x - 1$ is irreducible over Q.

    (v) $x^3 - 2$ is irreducible over Q.

    (vi) $x^3 + x^2 - 2x - 1$ is irreducible over Q.

    (vii) $x^4 + 2x + 2$ is irreducible over Q.

12. If a is rational and $(x - a)$ divides an integer monic polynomial, then prove that a must be an integer.

13. If $f(x) \in F[x]$, $F = Z_p$, p prime, and $f(x)$ is irreducible over F of degree n, prove that $\dfrac{F(x)}{f(x)}$ is a field with p^n elements.

14. If p is prime, prove that the polynomial $x^n - p$ is irreducible over Q.

15. Prove that the polynomial $1 + x + ... + x^{p-1}$, p prime is irreducible over Q.

    [*Hint:* Consider the polynomial $1 + (x+1) + ... + (x+1)^{p-1}$ and use the Einstein criterion].

CHAPTER 6

Vector Spaces

6.1 VECTOR SPACES

Definition

A vector space (or linear space) over a field K is a set X with mappings $(x, y) \to x + y$ of $X \times X$ into X, called addition and $(\lambda, x) \to \lambda x$ of $K \times X$ into X, called scalar multiplication such that the following axioms are satisfied; for x, y and z in X and λ, μ in K:

1. $(x+y)+z = x+(y+z)$.
2. $x+y = y+x$.
3. There exists an element $0 \in X$, called the zero vector such that $x+0=x$.
4. For each $x \in X$ there exists an element $(-x) \in X$ called the additive inverse or the negative of x such that $x+(-x)=0$
5. $\lambda(x+y) = \lambda x + \lambda y$
6. $(\lambda+\mu)x = \lambda x + \mu x$
7. $\lambda(\mu x) = (\lambda\mu)x$
8. $1.\ x = x$

The elements of X are called vectors and those of the field K are called scalars. A vector space X is called a real or complex vector space according as the field K is R, the field of real numbers, or C, the field of complex numbers.

Example 6.1

1. Any field is a vector space over itself.
2. For any $n \in N$, $R^n = \{(x_1, x_2, \cdots, x_n) : x_i \in R,\ i = 1, 2, \cdots, n\}$ is a real vector space with respect to addition and scalar multiplication defined as follows:

   $x + y = (x_1 + y_1, \cdots, x_n + y_n),\ \lambda x = (\lambda x_1, \cdots, \lambda x_n)$

where $x=(x_1,\cdots,x_n)$, $y=(y_1,\cdots,y_n)$ and $\lambda \in R$.

Similarly, $C^n=\{(z_1,z_2,\cdots,z_n): z_i \in C,\ i=1,2,\cdots,n\}$ is a complex vector space.

3. Let S denote the set of all sequences $x=\{x_n\}$ of real (complex) numbers. Then S is a real (complex) vector space under coordinate wise linear operations defined as follows:

   $\{x_n\}+\{y_n\}=\{x_n+y_n\}$, $\lambda\{x_n\}=\{\lambda x_n\}$

4. Let X be any nonempty set and $I(X)$ denote the set of all real valued (complex valued) functions defined on X. Then $I(X)$ is a real (complex) vector space with respect to pointwise linear operations defined as follows:

   $(f_1+f_2)(x)=f_1(x)+f_2(x)$

   $(\lambda f)(x)=\lambda f(x)$ for all $x \in X$

   Where $f_1, f_2 \in I(X)$ and λ is real or complex.

5. $M_{m,n}$ the set of all $m \times n$ (with m and n fixed positive integers) matrices with real (or complex) entries. If $A=(a_{ij})$ and $B=(b_{ij})$ are two such matrices, then we define the sum $A+B$ to be the matrix $C=(c_{ij})$ with elements:

   $c_{ij}=a_{ij}+b_{ij}$

   Similarly if $A=(a_{ij})$ is such a matrix and λ is a real (complex) number, then we define the product of λ and A to be the $m \times n$ matrix obtained by multiplying λ to every element of A. This is called the scalar multiplication, that is, multiplication of the matrix A by the scalar λ and is denoted by λA. It is left to the reader to prove that the set of all $m \times n$ real (or complex) matrices (with m and n fixed) form a real (or complex) vector space with respect to the addition and scalar multiplication defined as above.

6.1.1 Subspace

Let X be a vector space over a field K. A subspace of X is a nonempty subset E of X which is a vector space by itself. Note that E is a subspace of X iff $x+y$ and λx are in E whenever x and y are in E and $\lambda \in K$. It should be noted that the set $\{0\}$ consisting of just the zero vector and the vector space X itself are subspaces of X. These two subspaces are called trivial and all others are called nontrivial subspaces of X. The reader should also note that the arbitrary intersection of any family of subspaces is again a subspace. The intersection is nonempty since each subspace contains the zero vector.

If A is subset of X the linear span (hull) of A is the intersection, E, of all subspaces of X that contain A. E is a subspace and the smallest subspace of X containing A. E is sometimes called the subspace of X generated by A. Note that the linear span of the null set φ is $\{0\}$.

6.1.2 Quotient Space

Let X be a vector space over a field K and let M be a subspace of X. Define a relation $\sim$ in X as follows: for $x,\ y \in X$ if and only if $x-y \in M$ where $x-y=x+(-y)$. Then it is to see that $\sim$ is an equivalence relation and therefore divides X into disjoint equivalence classes. Denote the equivalence class of $x \in X$ by $x+M$ and the set of all equivalence classes by X/M. Define addition and scalar multiplication in X/M as follows: for $x, y \in X$ and $\lambda \in K$,

$$(x+M)+(y+M)=(x+y)+M$$
$$\lambda(x+M) = \lambda x+M$$

Then it is a routine verification that X/M is a vector space over K and the proof is left to the reader. (The vector space X/M is called the quotient space of X by M and the elements, $x + M$, are called cosets). We shall, however, only check that addition is well defined. Let $x_1+M=y_1+M$ and $x_2+M=y_2+M$. Then $x_1-y_1 \in M$, $x_2-y_2 \in M$. Since M is a subspace, $(x_1+x_2)-(y_1+y_2)=(x_1-y_1)+(x_2-y_2) \in M$ showing that $(x_1+x_2)+M=(y_1+y_2)+M$.

Definition

Let X be a vector space over a field K and let $\lambda_1, \lambda_2, \cdots, \lambda_n \in K$. An element $\lambda_1 x_1+\lambda_2 x_2+\cdots+\lambda_n x_n$ is called a linear combination of the elements $x_1, x_2, \cdots, x_n$ of X. It is also called a linear combination of the set $A=\{x_1, x_2, \cdots, x_n\}$. We write this element as $\sum_{i=1}^{n} \lambda_i x_i$.

It should be noted that the linear span of the set A is the set of all linear combinations of A. A subset A of X is called linearly independent if for every nonempty finite subset $\{x_1, x_2, \cdots, x_n\}$ of A the relation $\sum_{i=1}^{n} \lambda_i x_i = 0$ implies $\lambda_i=0$ for $i=1, 2, \cdots, n$.

Note that by this definition, the empty set is linearly independent. A subset A of X which is not linearly independent is called linearly dependent.

Observe that any subset of X containing the zero vector, in particular the set $\{0\}$, is linearly dependent.

A linearly independent subset of X which is maximal (with respect to set inclusion) is called a basis (Hamel basis) of X.

It is a basic question whether every vector space has a basis and we need the help of Zorn's lemma to answer this question.

6.1.3 Zorn's Lemma

In the study of analysis, algebra and topology one frequently faces situations in which tools of set theory are too weak to permit construction and proofs. We need an innocent appearing but actually a very strong axiom known as Zorn's lemma. It states that:

If X is a nonempty partially ordered set in which every chain (totally ordered set) has an upper bound in X then X possesses a maximal element.

There are several equivalent forms of Zorn's lemma, one of them is called the axiom of choice which was postulated by the German mathematician Ernst Zermelo. We only state this axiom and will not attempt to prove the equivalence. It states that:

The Cartesian product of nonempty family of nonempty sets is nonempty. In other words: if $\{X_i\}$ is a family of nonempty sets indexed by a nonempty set I then there exists a family $\{X_i\} i \in I$ such that $x_i \in X_i$ for each $i \in I$.

Theorem 6.1

Let X be a nonzero vector space, L a linearly independent subset of X. Then there exists a basis B of X such that $L \subset B$.

Proof: Let A denote the family of linearly independent subsets of X containing L. Clearly $A \neq \varphi$ since $L \in A$. Particularly order A by set inclusion. If $\mathcal{F}$ is any totally ordered subset of A, let $F = \cup \{E : E \in \mathcal{F}\}$.

F is linearly independent because if $x_1, x_2, \cdots, x_n \in F$ then since $\mathcal{F}$ is totally ordered, there exists a linearly independent $E \in \mathcal{F}$ such that $x_1, x_2, \cdots, x_n \in E$. Also, since $L \subset F \subset X$, F is an upper bound for $\mathcal{F}$ in A. Hence by Zorn's lemma, A contains a maximal element B. B is linearly independent and

$L \subset B \subset X$, since $B \in A$. It remains to show that span $(B) = X$. B being maximal in A, $B \neq \varphi$ and hence $0 \in$ span (B). Assume, to the contrary, that there exists $x \in X$ such that $x \in$ span (B). Then $x \neq 0$. In fact, $B \cup \{x\}$ is linearly independent. If $kx = 0$ with $k \in K$, then $k = 0$ since $x \neq 0$. If $kx + k_1 b_1 + \cdots + k_n b_n = 0$, with $k_1, k_2, \cdots, k_n \in K$ and $b_1, b_2, \cdots, b_n \in B$ then $kx = \sum_{i=1}^{n} k_i b_i$ and since $x \in$ span (B), it follows that $k = 0$.

Hence, $k_1 b_1 + k_2 b_2 + \cdots + k_n b_n = 0$ which implies that $k_1 = 0 = k_2 = \cdots = k_n$, by the linear independence of B. Thus we see that $B \cup \{x\}$ is in A, which properly contains B. This contradicts the maximality of B in A. Hence, span $(B) = X$ and this completes the proof.

Corollary 1

Every nontrivial vector space X has a basis.

Proof: Take $L = \varphi$ in above Theorem 6.1.

Corollary 2

If Y is a subspace of a vector space X then every basis of Y can be extended to a basis of X.

Proof: Let B be a basis for Y. Take $L = B'$ in the above Theorem 6.1.

The following theorem answers another fundamental question that if a vector space X has a finite set as a basis, then every basis for X has the same number of elements.

Theorem 6.2

If $S = \{v_1, v_2, \cdots, v_n\}$ is a basis of X and if $B = \{w_1, w_2, \cdots, w_m\} \subset X$ is linearly independent then $m \leq n$.

Proof: Since S is a basis for X every element of X thus in particular w_m is a linear combination of elements of S. Therefore $\{w_m, v_1, v_2, \cdots, v_n\}$ is linearly dependent. More over this set spans X since S does so. Thus some proper subset of these, say $w_m, v_{i1}, v_{i2}, \cdots, v_{ik}$ with $k \leq n-1$ forms a basis for X. Now repeat the procedure with the set $\{w_{m-1}, w_m, v_{i1}, v_{i2}, \cdots, v_{ik}\}$. From this linearly dependent set we can extract a basis of the form $w_{m-1}, w_m, v_{j1}, v_{j2}, \cdots, v_{js}$,

$s \leq n-2$. continuing in this way we get down to a basis of X of the form $w_2, \cdots, w_{m-1}, v_\alpha, v_\beta, \cdots$. Since, w_i is not a linear combination of $w_2, \cdots, w_{m-1}$, the above basis must include some v. To get to this basis we have introduced $m-1$ w's, each such introduction having cost us at least one v and yet there is a v left. Thus $m-1 \leq n-1$. Hence, $m \leq n$ and this completes the proof.

Corollary 1

If X has a basis consisting of n elements, then every set of m vectors with $m > n$ is linearly dependent.

Proof: Suppose $B = \{v_1, v_2, \cdots, v_n\}$ is a basis for X. Let $A = \{u_1, u_2, \cdots, u_m\}$ be a set of m vectors with $m > n$. If A is linearly independent, then by the above theorem $m > n$. This contradicts the fact that $m > n$ and completes the proof.

Corollary 2

If X has a basis consisting of n elements, then every other basis for X has also n elements.

Proof: Suppose $B_1 = \{v_1, v_2, \cdots, v_n\}$ and $B_2 = \{w_1, w_2, \cdots, w_m\}$ are two bases for X. Then B_1 and B_2 are linearly independent, Since, B_1 is a basis and B_2 is linearly independent, we have $m \leq n$. Now reversing the roles of B_1 and B_2 we get $n \leq m$. Thus $m = n$ and this completes the proof.

The above result enables us to introduce the following definition.

Definition

A vector space is said to have dimension n if number of elements of any basis for it is n. If the dimension of X equals n for some $n \in N$ then X is said to be finite dimensional. If $X = \{0\}$, then its dimension is defined to be zero and in this case also X is said to be finite dimensional. If X not finite dimensional, it is called infinite dimensional.

Example 6.2

1. Let $X = R^3$ and let

   $e_1 = (1, 0, 0) = i$

   $e_2 = (0, 1, 0) = j$

   $e_3 = (0, 0, 1) = k$

Then the set $\{e_1, e_2, e_3\}$ is linearly independent. Moreover, it is a basis for $R^3$and R^3 has dimension 3. This basis is called the standard basis for R^3. On the other hand $\{(1, 0, 1), (1, 1, 0), (-1, 0, -1)\}$ is linearly dependent and hence it is not a basis.

2. Let $X = R^n$. Let e_i be the element of R^n whose i^{th} coordinate is 1 and all others are zeros. Then the set $\{e_1, e_2, \cdots, e_n\}$ is linearly independent and forms a basis for R^n.

   This is known as the standard basis for R^n.

Definition

Let X be a linear space over the complex field C. An inner product on X is a function $(,)$: $X \times X \rightarrow C$ which satisfies the following conditions:

1. $(x, y) = \overline{(y, x)}$ for $x, y \in X$
2. $(\lambda x + \mu\, y, z) = \lambda(x, z) + \mu(y, z)$ for $\lambda, \mu \in C; x, y, z \in X$
3. $(x, x) \geq 0, (x, x) = 0$ iff $x = 0$.

An (complex) inner product space X is a linear space over C with an inner product defined on it. A (real) inner product space is similarly defined.

From (i) and (ii) it follows that for an inner product space X.

$(x,\ \lambda y + \mu z) = \bar{\lambda}(x, y) + \bar{\mu}(x, z)$ for any $x, y, z \in X$ and $\lambda, \mu \in C$ where $\bar{\lambda}$ throughout denotes the complex conjugate of λ in C. Unless otherwise stated an inner product space will mean a complex inner product space.

Example 6.3

1. R^n is a (real) inner product space with the inner product,

   $$(x, y) = \sum_{k=1}^{n} x_k y_k$$

2. C^n is a (complex) inner product space with,

   $$(x, y) = \sum_{k-1}^{n} x_k \overline{y_k}$$

   If X is an inner product space then for $x \in X$, we define $\|x\| = (x, x)^{1/2}$. For $x, y \in X$ we define $d(x, y) = \|x - y\|$.

Theorem 6.3 (Cauchy –Schwarz inequality)

If X is an inner product space and $x, y \in X$ then $|(x, y)| \leq \|x\| \ \|y\|$.

Proof: If $y = 0$, then equality holds. Let $y \neq 0$ and let $\lambda \in C$. Then we have

$$\begin{aligned} 0 \leq \|x - \lambda y\|^2 &= (x - \lambda y, x - \lambda y) \\ &= (x, x) - \lambda(y, x) - \bar{\lambda}(x, y) + |\lambda|^2 (y, y) \\ &= \|x\|^2 - \lambda(y, x) - \bar{\lambda}(x, y) + |\lambda|^2 \|y\|^2 \end{aligned}$$

Put $\lambda = \dfrac{(x, y)}{\|y\|^2}$. Then

$$0 \leq \|x\|^2 - \frac{(x, y) \ (y, x)}{\|y\|^2} - \frac{\overline{(x, y)} \ (x, y)}{\|y\|^2} + \frac{\|(x, y)\|^2}{\|y\|^2} = \|x\|^2 - \frac{|(x, y)|^2}{\|y\|^2}$$

Thus,

$$|(x, y)| \leq \|x\| \ \|y\|.$$

Theorem 6.4

If X is an inner product space then,

1. $\|x\| = 0$ if and only if $x = 0$
2. $\|\lambda x\| = |\lambda| \|x\|$ for $\lambda \in C$ and $x \in X$
3. $\|x + y\| \leq \|x\| + \|y\|$ for $x, y \in X$.

Proof: Clearly $\|x\| = 0$ if and only if $x = 0$. For $x \in X$ and $\lambda \in C$ we have,

$$\|\lambda x\| = (\lambda x, \lambda x)^{1/2} = (\lambda \bar{\lambda}(x, x))^{1/2} = |\lambda| \ \|x\|$$

If $x, y \in X$ by using Cauchy-Schwarz inequality we have,

$$\begin{aligned} \|x + y\|^2 &= (x + y, x + y) = (x, x) + (x, y) + (y, x) + (y, y) \\ &= \|x\|^2 + \|y\|^2 + 2 \ re(x, y) \\ &\leq \|x\|^2 + \|y\|^2 + 2 \ |(x, y)| \\ &\leq \|x\|^2 + \|y\|^2 + 2 \ \|x\| \ \|y\| \\ &= (\ \|x\| + \|y\|)^2 \end{aligned}$$

and this completes the proof.

Proposition 6.1

If X and Y are in an inner product space, then

$$4(x,y)=\|x+y\|^2-\|x-y\|^2+i\ \|x+iy\|^2-i\ \|x-iy\|^2$$

Proof: This follows from direct computation.

Proposition 6.2

If X and Y are in an inner product space, then

$$\|x+y\|^2+\|x-y\|^2=2\left(\|x\|^2+\ \|y\|^2\right)$$

Proof: This follows from direct computation. This is known as Parallelogram law.

Proposition 6.2

If X is an inner product space, then

1. $d(x,y)=0$ iff $x=y$
2. $d(x,y)=d(y,x)$
3. $d(x,z)\le d(x,y)+d(y,z)$

Proof: Use theorem 6.4.

One of the most important geometrical properties is a notion of orthogonality. In this section we shall observe that such a concept is possible in an inner product space. In fact inner product spaces possess many interesting properties of the Euclidean space. It is possible to extend many theorems from Euclidean geometry to inner product spaces. We shall first introduce the definition of orthogonality and then prove an important result which is the familiar Pythagorean theorem.

Definition

Two elements x and y in an inner product space X are said to be orthogonal, denoted $x\perp y$, if $(x,y)=0$. A subset S of X is said to be orthogonal if $x\perp y$ for all $x,y\in S$. S is said to be orthonormal, if, in addition $\|x\|=1$ for every $x\in S$.

Theorem 6.5 (Pythagorean Theorem)

If $\{x_1, x_2, \cdots, x_n\}$ is an orthogonal subset of an inner product space, then

$$\left\|\sum_{i=1}^{n} x_i\right\|^2 = \sum_{i=1}^{n} \|x_i\|^2$$

Proof: We have,

$$\begin{aligned}\left\|\sum_{i=1}^{n} x_i\right\|^2 &= \left(\sum_{i=1}^{n} x_i, \sum_{j=1}^{n} x_j\right)\\ &= \sum_{i=1}^{n}(x_i, x_i) + \sum_{\substack{i,j=1\\ i\neq j}}^{n}(x_i, x_j)\\ &= \sum_{i=1}^{n}(x_i, x_i)\\ &= \sum_{i=1}^{n}\|x_i\|^2\end{aligned}$$

Theorem 6.6 (Gram-Schmidt Orthogonalisation)

Let $S = \{x_1, x_2, \cdots,\}$ be a linearly independent sequence in an inner product space. Then there exists an orthonormal sequence $T = \{y_1, y_2, \cdots,\}$ such that span $(S) =$ span (T).

Proof: Since S is linearly independent $x_i \neq 0$ for each k. Define

$$y_1 = \frac{x_1}{\|x_1\|}, \text{ so that } \|y_1\| = 1$$

Define $v = x_2 - (x_2, y_1)y_1$. Then $v \perp y_1$ and $v \neq 0$, since $\{x_1, x_2\}$ is linearly independent. Hence, $y_2 = \frac{x_2}{\|x_2\|}$ is orthogonal to y_1 and $\|y_2\| = 1$. We now inductively define $v = x_n - \sum_{k=1}^{n-1}(x_n, y_k)y_k$ and $y_n = \frac{v}{\|v\|}$. It is clear from the construction that span $(S) =$ span (T).

6.1.4 Discussions

If X is a nonzero inner product space, then the family of all orthonormal sets is non empty. This family is clearly a partially ordered set with respect to set with respect to set inclusion as the partial order. An orthonormal set $\{e_i\}$ in X is said to be complete if it is maximal in this partially ordered set, i.e., if it is impossible to adjoin a vector to $\{e\}$ such that $\{e_i, e\}$ is an orthonormal set which property contains $\{e_i\}$.

It should be noted that every nonzero inner product space X contains a complete orthonormal set. In the partially ordered set of all orthonormal sets the union of any chain of orthonormal sets is clearly an upper bound for the chain and hence the above statement follows at once from Zorn's lemma.

6.2 LINEAR MAPPINGS

Definition

Let X and Y be any two vector spaces over a field K. A mapping $f: X \to Y$ is called a linear mapping (linear transformation or homomorphism) if:

$$f(\lambda_1 x_1 + \lambda_2 x_2) = \lambda_1 f(x_1) + \lambda_2 f(x_2) \text{ for all } x_1, x_2 \in X \text{ and } \lambda_1, \lambda_2 \in K.$$

Further, if f is 1–1, then it is called an isomorphism and two vector spaces X and Y are said to be isomorphic if there exists an isomorphism from X onto Y.

A linear map f from X into itself is sometimes called a linear operator. If f is a linear map from X in to K then it is called a linear functional. Note that if f is a linear map from X in to Y then it maps the zero of X in to zero of Y.

Example 6.4

1. The zero mapping $0: X \to Y$ defined by,

   $0(x) = 0$ for all $x \in X$

   is a linear mapping. It is also called trivial mapping.
2. The identity map $id: X \to X$ defined by

   $id(x) = x$

   Is also a linear map.
3. If $X = R$ (note that R is a vector space over itself), then the function $f: R \to R$ defined by $f(x) = x + a$ (where $a \in R$ is fixed) is usually called a linear map. But it is not really so.
4. A constant map $f: R \to R$ defined by $f(x) = a$ (where $a \in R$ is fixed) is not a linear map.

   Two important subsets (in fact subspaces) are associated with a linear map $f: X \to Y$.

One is called the kernel or the null space of f and is denoted by $N(f)$, the other is the range space of f and is denoted by $R(f)$.

Definition

Let $f: X \to Y$ be a linear map. Then

$$N(f) = \{x \in X : f(x) = 0\}$$

and

$$R(f) = \{f(x) \in Y : x \in X\}$$

We have.

Theorem 6.7

Let $f: X \to Y$ be a linear map. Then

1. $R(f)$ is a subspace of Y
2. $N(f)$ is a subspace of X
3. f is 1–1, i.e. f is an isomorphism iff $N(f) = \{0\}$.

Proof:

1. Let $y_1, y_2 \in R(f)$. Then there exist $x_1, x_2 \in X$ such that $f(x_1) = y_1$ and $f(x_2) = y_2$. Let λ_1, λ_2 be some scalars. Then since f is linear,
   $$\begin{aligned} f(\lambda_1 x_1 + \lambda_2 x_2) &= \lambda_1 f(x_1) + \lambda_2 f(x_2) \\ &= \lambda_1 y_1 + \lambda_2 y_2 \end{aligned}$$
   But $\lambda_1 x_1 + \lambda_2 x_2 \in X$ since X is a vector space. Hence $\lambda_1 y_1 + \lambda_2 y_2$ being the image of an element of X in $R(f)$ is a subspace of Y.
2. Let $x_1, x_2 \in N(f)$ so that $f(x_1) = 0$ and $f(x_2) = 0$. Therefore for any scalars λ_1, λ_2 we have,
   $$\begin{aligned} f(\lambda_1 x_1 + \lambda_2 x_2) &= \lambda_1 f(x_1) + \lambda_2 f(x_2) \\ &= \lambda_1.0 + \lambda_2.0 = 0 \end{aligned}$$
   Hence, $\lambda_1 x_1 + \lambda_2 x_2 \in N(f)$ showing that $N(f)$ is a subspace of X.
3. First of all observe that if f is a linear map then $f(0) = 0$ and so $0 \in N(f)$. Now suppose that f is 1–1. Then $f(x_1) = f(x_2)$ implies $x_1 = x_2$. If $x \in N(f)$ then $f(x) = 0 = f(0)$ and hence $x = 0$. This shows that,

   $N(f) = \{0\}$.

Conversely, suppose $N(f)=\{0\}$. We have to show that f is 1–1. Let $f(x_1)=f(x_2)$. Then $0=f(x_1)-f(x_2)=f(x_1-x_2)$.

So, $x_1-x_2 \in N(f)=\{0\}$. Thus, $x_1-x_2=0$, *i.e.* $x_1=x_2$ which proves that f is 1–1.

Definition

If $R(f)$ is finite dimensional as vector space, then the dimension of $R(f)$ is called the rank of f. Similarly, if $N(f)$ is finite dimensional as a vector space, the dimension of $N(f)$ is called the nullity of f.

We shall now define two operations (one is called addition and the other scalar multiplication) on the set of all linear transformations from one vector space into another and show that this forms a vector space.

Let X and Y be vector spaces over the same field K. Let $L(X, Y)$ denote the set of all linear maps from X in to Y. Define addition and scalar multiplication in $L(X, Y)$ by,

$$(f_1+f_2)(x)=f_1(x)+f_2(x)$$
$$(\lambda f)(x)=\lambda(f(x))$$

Theorem 6.8

$L(X,Y)$ is a vector space under addition and scalar multiplication defined above.

Proof: We shall only show that $L(X,Y)$ is closed under addition and scalar multiplication. The checking of the other axioms is left to the reader as an exercise.

Let $f_1, f_2 \in L(X,Y)$ for all $x_1, x_2 \in X$ and $\lambda_1, \lambda_2 \in K$, we have

$$\begin{aligned}(f_1+f_2)(\lambda_1x_1+\lambda_2x_2) &= f_1(\lambda_1x_1+\lambda_2x_2)+f_2(\lambda_1x_1+\lambda_2x_2)\\ &= [\lambda_1f_1(x_1)+\lambda_2f_1(x_2)]+[\lambda_1f_2(x_1)+\lambda_2f_2(x_2)]\\ &= [\lambda_1f_1(x_1)+\lambda_1f_2(x_1)]+[\lambda_2f_1(x_2)+\lambda_2f_2(x_2)]\\ &= \lambda_1\ (f_1+f_2)(x_1)+\ \lambda_2\ (f_1+f_2)(x_2)\end{aligned}$$

Thus, $f_1+f_2 \in L(X,Y)$.

Now, let $f \in L(X,Y)$ and $\lambda \in K$.

Then,

$$\begin{aligned}(\lambda f)(\lambda_1 x_1 + \lambda_2 x_2) &= \lambda \quad f(\lambda_1 x_1 + \lambda_2 x_2)\\ &= \lambda \quad [\lambda_1 f(x_1) + \lambda_2 f(x_2)]\\ &= \lambda_1 \quad \lambda f(x_1) + \lambda_2 \quad \lambda f(x_2)\\ &= \lambda_1 \quad (\lambda f)(x_1) + \lambda_2 \quad (\lambda f)(x_2)\end{aligned}$$

This shows that $\lambda f \in L(X, Y)$.

Remark

As a particular case of the above result it follows that $L(X) = L(X, X)$, the set of all linear operators defined on a vector space *X*, is a vector space.

Also *L*(*X, K*), the set of all linear functionals defined on *X*, is a vector space, this is called the algebraic dual of *X* and is written as X^*or X^+.

Let *X* and *Y* be any two vector spaces over a field *K* of dimensions *m* and *n* respectively. Let $(\alpha_1, \alpha_2, \cdots, \alpha_m)$ and $(\beta_1, \beta_2, \cdots, \beta_n)$ be the bases of *X* and *Y* respectively. Then any linear transformation from *X* in to *Y* is precisely an $m \times n$ matrix over the field *K*. To show this, let *f* be a linear transformation from *X* into *Y*. Observe that each $\alpha_i \in X$, so that $f(\alpha_i) \in Y$. Since $\{\beta_i\}$ is a basis for *Y* each $f(\alpha_i)$ can be expressed as a linear combination of β_i's. Hence, we have

$$f(\alpha_i) = \sum_{j=1}^{n} \alpha_{ij} \beta_j, \quad \alpha_{ij} \in K$$

Thus, $A = (\alpha_{ij})$ is an $m \times n$ matrix over *K*. Then relative to two given bases $(\alpha_1, \alpha_2, \cdots, \alpha_m)$ and $(\beta_1, \beta_2, \cdots, \beta_n)$ of *X* and *Y*, there is a linear transformation *f* defined by,

$$f(\alpha_i) = \sum_{j=1}^{n} \alpha_{ij} \beta_j$$

Thus relative to the choice of fixed bases in *X* and *Y* any linear transformation of *X* in to *Y* can be represented as an $m \times n$ matrix over the field *K* where *m* and *n* are the dimensions of *X* and *Y* respectively.

Let *X* and *Y* be as above. Let *f* and *g* be any two linear transformations of *X* into *Y*.

Then,

$$f(\alpha_i)=\sum_j a_{ij}\beta_j,\ a_{ij}\in K$$

$$g(\alpha_i)=\sum_j b_{ij}\beta_j,\ b_{ij}\in K$$

where $A=(a_{ij})$ and $B=(b_{ij})$ are $m\times n$ matrices over K.

Now,

$$\begin{aligned}(f+g)(\alpha_i)&=f(\alpha_i)+g(\alpha_i)\\&=\sum_j a_{ij}\beta_j+\sum_j b_{ij}\beta_j\\&=\sum_j (a_{ij}+b_{ij})\beta_j\end{aligned}$$

Further if $\lambda\in K$ then

$$\begin{aligned}(\lambda f)(\alpha_i)&=\lambda\ f(\alpha_i)\\&=\lambda\sum_j a_{ij}\beta_j\\&=\sum_j (\lambda a_{ij})\beta_j\end{aligned}$$

Because of the above facts and by virtue of theorem (6.8) we have the following result.

The set of all $m\times n$ matrices over a field K is a vector space over K of dimension mn with respect to addition of matrices and multiplication of a matrix by a scalar defined as follows.

If $A=(a_{ij})$, $B=(b_{ij})$ are $m\times n$ matrices over K and if $\lambda\in K$ then,

$$A+B=(a_{ij}+b_{ij})$$
$$\lambda A=(\lambda a_{ij})$$

Let X, Y and Z be any three vector spaces over the same field K. Let $f:X\to Y$ and $g:Y\to Z$ be any two linear transformations. Then the composition $gf:X\to Z$ defined by,

$$(gf)(x)=g\big(f(x)\big)\ \text{for all } x\in X$$

Then it is to see that gf is a linear transformation from X to Z. For if $x_1,x_2\in X$ and $\lambda\in K$ we have,

$$\begin{aligned}(gf)(x_1+x_2) &= g\left(f(x_1+x_2)\right)\\ &= g\left(f(x_1)\right)+g\left(f(x_2)\right)\\ &= (gf)(x_1)+(gf)(x_2)\end{aligned}$$

and

$$\begin{aligned}(gf)(\lambda x) &= g\left(f(\lambda x)\right) = g\left(\lambda f(x)\right)\\ &= \lambda\left[g\left(f(x)\right)\right] = \lambda\left[(gf)(x)\right]\end{aligned}$$

Thus the composition of two linear maps f and g is a linear map provided the composition is defined and it is defined only when the range of f is contained in the domain g.

Let X, Y and Z be vector spaces over K with dimensions m, n and p respectively. Let $(\alpha_1, \alpha_2, \cdots, \alpha_m)$, $(\beta_1, \beta_2, \cdots, \beta_n)$ and $(\gamma_1, \gamma_2, \cdots, \gamma_p)$ be three bases of X, Y and Z respectively. We write

$$f(\alpha_i) = \sum_j a_{ij}\beta_j$$

$$g(\beta_j) = \sum_k b_{jk}\gamma_k$$

So that,

$$A = (a_{ij}),\ B = (b_{jk})$$

are the two $m \times n$ and $n \times p$ matrices which represent the linear transformations f and g respectively relative to the given bases.

Then we have,

$$\begin{aligned}(gf)(\alpha_i) &= g\left(f(\alpha_i)\right)\\ &= \sum_j g\left(a_{ij}\beta_j\right)\\ &= \sum_j a_{ij}\ \ g\left(\beta_j\right)\\ &= \sum_j a_{ij}\ \ \sum_k b_{jk}\ \ \gamma_k\\ &= \sum_k\ \sum_j a_{ij}b_{jk}\gamma_k\end{aligned}$$

Now the number of columns of A is same as the number of rows of B. Hence the product of two matrices $A = (a_{ij})$, $B = (b_{jk})$ is defined to be the $m \times p$ matrix $C = (c_{ik})$ where,

$$c_{ik}=\sum_k \sum_j a_{ij}\, b_{jk}$$

Therefore, it follows that the product matrix C represents the composition gf of the linear transformation f and g.

1. $f_1(g_1+g_2)=f_1g_1+f_1g_2$
2. $(f_1+f_2)g_1=f_1g_1+f_2g_1$
3. $h\ (f_1g_1)=(hf_1)g_1$
4. $(\alpha f_1)g_1=\alpha(f_1g_1)$
5. $id_y f_1=f_1,\ f_1id_x=f_1$

Proof: We shall prove only (i). The proofs of other parts are left to the reader as an exercise.

Let $x\in X$.. Then

$$\begin{aligned}(f_1(g_1+g_2))(u) &= f_1((g_1+g_2)(u))\\ &= f_1(g_1(u)+g_2(u))\\ &= f_1(g_1(u))+f_1(g_2(u))\\ &= (f_1g_1)(u)+(f_1g_2)(u)\end{aligned}$$

and this completes the proof of (i).

Let A, B, C, D be matrices such that the multiplication stated below are defined. Then

1. $A(B+C)=AB+AC$
2. $(B+C)D=BD+CD$
3. $A(BC)=(AB)C$
4. $(\alpha A)B=\alpha(AB)$ where α is a scalar
5. $IA=AI=A$ where I is the identity matrix
6. $AB\neq BA$
7. $AB=0$ does not imply that $A=0$ or $B=0$.

where 0 denotes the zero matrix.

Proof: The proofs (1)–(5) are left to the reader as an exercise. For the proof of (6) observe that if:

$$A=\begin{pmatrix}1 & 0\\ 0 & 0\end{pmatrix},\quad B=\begin{pmatrix}0 & 1\\ 1 & 0\end{pmatrix}$$

Then,

$$AB = \begin{pmatrix} 0 & 1 \\ 0 & 0 \end{pmatrix}, \quad BA = \begin{pmatrix} 0 & 0 \\ 1 & 0 \end{pmatrix}$$

To prove (7) take,

$$A = \begin{pmatrix} 1 & 1 \\ 2 & 2 \end{pmatrix}, \quad B = \begin{pmatrix} -1 & 1 \\ 1 & -1 \end{pmatrix}$$

Then

$$AB = \begin{pmatrix} 0 & 0 \\ 0 & 0 \end{pmatrix}$$

but neither A nor B is a zero matrix.

Let A and B be any two sets and let $f : A \to B$ be a bijection. We have discussed in Chapter-1, that in this case the inverse of f written as f^{-1}exists and is defined as a map $f^{-1} : B \to A$ by $f^{-1}(y) =$ unique $x \in A$ for which $f(x) = y$. Further $f^{-1} : B \to A$ is also bijective. We shall now observe that bijective linear maps from one vector space in to another play a vital role in the study of linear transformations.

Definition

Let X and Y be vector spaces over K. A linear map $f : X \to Y$ is said to be nonsingular (invertible) if it is bijective. Note that such a mapping is also called an isomorphism from X on to Y and in this case X and Y are said to be isomorphic. Observe that a linear map $f : X \to Y$ is nonsingular iff it has an inverse, which is denoted by f^{-1}.

Example 6.5

1. Let X be any vector space over K. Let the identity map $id_x : X \to X$ be defined by $id_x : X \to X$ for all $x \in X$. Then id_x^{-1} exists and $id_X^{-1} = id_X$.
2. Let $f : R^2 \to R^2$ be defined by

   $f(x_1, x_2) = (x_1, -x_2)$

   Then f is bijective and $f^{-1}(y_1, y_2) = (y_1, -y_2)$ for

   $f(y_1, y_2) = (y_1, -(-y_2)) = (y_1, y_2)$

3. Let S denote the set of all sequences of real numbers. We have noted earlier that S is a vector space over the field R of real numbers with regard to addition and scalar multiplication defined as follows:
   For $x=(x_1, x_2, \cdots, x_n, \cdots)$, $y=(y_1, y_2, \cdots, y_n, \cdots)$ and $\alpha \in R$
   $x+y=(x_1+y_1, x_2+y_2, \cdots, x_n+y_n, \cdots)$
   $\alpha x=(\alpha x_1, \alpha x_2, \cdots, \alpha x_n, \cdots)$
   Define $f: S \to S$ by
   $$f(x)=f(\{x_1, x_2, \cdots, x_n, \cdots\})$$
   $$=\{x_2, \cdots, x_n, \cdots\}$$
   In this case f is linear $R(f)=S$, f is onto. But f is not 1–1 and $N(f)$ is the set of all sequences of the form $\{z, 0, 0, \cdots\}$. This mapping f does not have an inverse.
4. Let S be as above. Define $g: S \to S$ by,
   $$g(x)=g(\{x_1, x_2, \cdots, x_n, \cdots\})=\{0, x_1, x_2, \cdots\}$$
   Hence g is linear, g is 1–1, $N(f)=\{0, 0, \cdots\}$. But g is not on to, $R(f) \neq S$ since the sequence $\{1, 1, \cdots\}$. has no pre-image in S under g and hence g does not have an inverse.

Theorem 6.11

Let X and Y be vector spaces over a field K and let $f: X \to Y$ be an invertible linear map. Then $f^{-1}: Y \to X$ is linear and bijective.

Proof: We shall only prove the linearity of f^{-1}. For this let $y_1, y_2 \in Y$. Let $f^{-1}(y_1)=x_1$ and $f^{-1}(y_2)=x_2$. We now have $f(x_1)=y_1$ and $f(x_2)=y_2$.

Hence,

$y_1+y_2=f(x_1)+f(x_2)=f(x_1+x_2)$ since f is linear and so

$$f^{-1}(y_1+y_2)=x_1+x_2=f^{-1}(y_1)+f^{-1}(y_2)$$

Similarly for $\lambda \in R$

$$\lambda y_1=\lambda f(x_1)=f(\lambda x_1)$$

and hence $f^{-1}(\lambda y_1)=\lambda x_1=\lambda f^{-1}(y_1)$

and this completes the proof.

Theorem 6.12

Let X and Y be vector spaces over a field K and let $f: X \to Y$ be a linear map. Then f is invertible iff there exists a linear map $g: Y \to X$ such that $fg = id_y$ and $gf = id_x$ and this mapping g is written as f^{-1}.

Proof: If f is invertible, then clearly there exists $g = f^{-1}$ with the stated properties.

Conversely, let f and g exist with the property that $fg = id_y$ and $gf = id_x$.

Let $x \in N(f)$. Then $(gf)(x) = id_x(x) = x$ $(gf)(x) = g(f(x)) = g(0) = 0$.

Thus $N(f) = \{0\}$ and hence f is invertible and completes the proof.

Theorem 6.13

Let X, Y and Z be vector spaces over a field K. Let $f: X \to Y$ and $g: Y \to Z$ be linear maps. If f and g are invertible, then fg is invertible and $(fg)^{-1} = g^{-1} f^{-1}$.

Proof: Let f and g have inverse f^{-1} and g^{-1} respectively. We now have

$$f\ f^{-1} = id_Y,\ f^{-1}f = id_X,\ gg^{-1} = id_z,\ g^{-1}g = id_Y$$

Further,

$$\begin{aligned}(fg)\left(g^{-1}f^{-1}\right) &= f\left(gg^{-1}f^{-1}\right) = f\left(id_Z f^{-1}\right)\\ &= f\ f^{-1} = id_Y\\ \left(g^{-1}f^{-1}\right)(fg) &= g^{-1}\left(f^{-1}fg\right) = g^{-1}\left(id_X g\right)\\ &= g^{-1}g = id_Y\end{aligned}$$

Therefore, fg is invertible and $(fg)^{-1} = g^{-1}f^{-1}$.

Let X be a vector space over the field K of dimension n and let a fixed basis be chosen for it. Let f be a linear mapping from X in to X. Then relative to the given basis f will generate an $n \times n$ matrix (a matrix with n rows and n columns). Such a matrix (a matrix having same number of rows and columns) is called a square matrix. An $n \times n$ square matrix is called a square matrix of order n. A square matrix $A = (a_{ij})$ of order n such that $a_{ij} = 1 (i = j), \ = 0 (i \neq j)$ is called a unit matrix or identity matrix of order n.

Just as a linear operator relative to a given basis generates a square matrix of order n, a square matrix of order n also defines a linear operator on X.

A square matrix A of order n is said to be invertible (nonsingular) if there exists a square matrix B of the same order such that

$$AB = BA = I_n$$

where I_n denotes the identity matrix of order n. In this case B is called an inverse of A. But it is easy to observe that if an inverse of a square matrix exists, then it is unique. Hence we call the inverse and write A^{-1} for the inverse of A if it exists. It is also easy to observe that $\left(A^{-1}\right)^{-1} = A$.

Theorem 6.14

A square matrix A of order n is invertible iff the corresponding linear operator f (with regard to some fixed basis) is invertible.

Proof: Let A be invertible. Then there exists a square matrix B of order n such that $AB = I_n = BA$ where I_n denotes the identity matrix of order n. Now there exists a linear operator g which corresponds to B with regard to the given fixed basis with the property that $fg = I = gf$ where I stands for the identity operator. Therefore f is invertible.

Conversely, let f be invertible. Then there exists a linear operator such that $fg = I = gf$. Let B be the matrix associated with g with regard to the given fixed basis. Then $AB = I_n = BA$. Thus, A is invertible and this completes the proof of the theorem.

Theorem 6.15

If A and B are any two nonsingular (invertible) square matrices of order n, then AB is also invertible and $(AB)^{-1} = B^{-1}A^{-1}$.

The proof of this is left to the reader as an exercise.

Let X and Y be any two vector spaces over a field K of dimensions m and n respectively. Let $(\alpha_1, \alpha_2, \cdots, \alpha_n)$ and $(\beta_1, \beta_2, \cdots, \beta_n)$ be the fixed bases of X and Y respectively. Then we have already observed that any linear transformation $f: X \to Y$ relative to the given fixed bases can be represented as an $n \times n$ matrix $A = (a_{ij})$ where,

$$f(\alpha_i) = \sum_{j=1}^{n} a_{ij}\beta_j, \quad a_{ij} \in K \quad (i = 1, 2, 3, \dots, m).$$

A is a matrix with m rows and n columns and represents the linear transformation f from the m dimensional space X into the n dimensional space Y. Let us now form a new matrix from the matrix A by changing the rows into columns and columns into rows. This new matrix so obtained is called the transpose of A and is written as A^T. Observe that A^T is an $n \times m$ matrix i.e. a matrix having n rows and m columns. Therefore A^T must represent a linear transformation from an n dimensional vector space into a space of dimension m. The linear transformation associated with A^T is called the transpose of the mapping f and is denoted f^T.

Since, f is a linear map from an n dimensional space into an m dimensional space, one may expect that it is from Y into X. But it is not really so, however f^T is a map from Y^* into X^*.

The following theorem is left to the reader as an exercise.

Theorem 6.16

1. Let A and B be any two $m \times n$ matrices and α a scalar. Then

   (a) $(A+B)^T = A^T + B^T$

   (b) $(\alpha A)^T = \alpha A^T$

   (c) $(A^T)^T = A.$

2. Let A and B respectively be $m \times n$ and $n \times p$ matrices. Then

   $(AB)^T = B^T A^T$

3. If A is a nonsingular square matrix then A^T is also nonsingular and

   $(A^{-1})^T = (A^T)^{-1}$

Let A be an $m \times n$ matrix over a field K. Let f be the linear transformation associated with the matrix A with regard to fixed given bases from a vector space X of dimension m into an n dimensional space Y over K. We have defined $R(f)$, the range space of f and the rank of f as the dimension of $R(f)$. We now define the rank of the matrix A as the dimension of the range space $R(f)$ of the associated linear transformation f. The concept of rank of a matrix is useful, in the study of simultaneous linear equations and it can be defined in several other equivalence ways. Since this book is designed to be a text on abstract algebra and not on Linear algebra we shall not go into the details of matrix theory. We close this section with the statement of the following theorem which the reader should try to prove.

Theorem 6.17

The transpose of a matrix A has the same rank as that of A.

6.3 MODULES

The concept of a module is a generalization of that of a vector space. We define vector spaces over a field while a module is defined on an arbitrary ring.

Definitions

A module over a ring R (or an R-module) is a nonempty set M with mappings $(x, y) \to x + y$ of $M \times M$ into M called addition and $(\lambda, x) \to \lambda x$ of $R \times M$ into M called scalar multiplication such that,

1. M is an additive abelian group
2. For $x, y \in M$ and $\lambda, \mu \in R$

   (a) $\lambda(x + y) = \lambda x + \lambda y$

   (b) $(\lambda + \mu)x = \lambda x + \mu x$

   (c) $\lambda(\mu x) = (\lambda\mu)x$ $\lambda(\mu x) = (\lambda\mu)x$

Example 6.6

1. Any ring R is a module over itself.
2. A vector space X over a field K is a module over K.
3. Any abelian group G is a module over the ring Z of integers.

Definition

Let M be a module over a ring R. A submodule of M is a nonempty subset N of M such that

1. N is an additive subgroup of M.
2. $\lambda \in R$ and $x \in N$ imply that $\lambda x \in N$.

In other words, a nonempty subset N of an R module M is an R -submodule if

1. $x_1, x_2 \in N$ imply $x_1 - x_2 \in N$
2. $\lambda \in R,\ x \in N$ imply $\lambda x \in N$.

Example 6.7

1. If a ring R is regarded as an R -module, then any ideal of R is a submodule of R.
2. If X is a vector space over a field K then any subspace of X is a K submodule of X.
3. If L and N are submodules of a module M then $L \cap N$ is a submodule of M.
4. If L and N are submodules of a module M then $L+N$ is a submodule of M where $L+N=\{l+n : l \in L \text{ and } n \in N\}$.

6.3.1 Quotient Module

Let M be an R-module and let N be an R-submodule of M. Define a relation in M as follows: for $x, y \in M$, $x - y$ iff $x - y \in N$. This is an equivalence relation and hence divides M into disjoint equivalence classes. Denote the equivalence class of $x \in M$ by $x+N$ and the set of all such equivalence classes by M/N. Define addition and scalar multiplication in M/N by

$$(x+N)+(y+N)=(x+y)+N$$
$$\lambda(x+N) = \lambda x + N$$

where $x, y \in M$ and $\lambda \in R$. Then observe that M/N is an R-module under addition and scalar multiplication defined above. The module M/N is called the quotient module or the factor module of M with respect to N. The reader should check the details and observe that the quotient modules are constructed in a similar way we constructed quotient groups, rings and quotient spaces.

Definition

Let M and N be two R-modules. A mapping $T: M \rightarrow N$ is called a homomorphism (or R-homomorphism or module homomorphism) if for $m_1, m_2, m \in M$ and $\lambda \in R$.

$$T(m_1 + m_2) = T(m_1) + T(m_2)$$
$$T(\lambda m) = \lambda T(m).$$

A homomorphism T is said to be an isomorphism if T is 1–1. Two R-modules M and N are said to be isomorphic if there exists an isomorphism from M onto N. As in the case of vector spaces two important subsets are associated with a homomorphism T: the kernel of T and the image of T. If T is a homomorphism from an R-module M into an R-module N then the kernel of T, $K(T)$ is defined by

$$K(T)=\{x\in M:\ T(x)=0\}$$

and the image of T, $I(T)$ is defined by

$$I(T)=\{T(x):x\ \in M\}.$$

We have.

Theorem 6.18

Let M and N be two R-modules and T a homomorphism from M into N. Then

1. $K(T)$ is a R-submodule of M.
2. $I(T)$ is a R-submodule of N.
3. T is an isomorphism iff $K(T) = \{0\}$.

This is left to the reader as an exercise.

EXERCISE

1. Prove that the intersections of any two subspaces of a vector space is a subspace. Is the union of two subspaces of a vector space a subspace?
2. If U and W are subspaces of a vector space V then show that $U+W=\{v\in V:v=u+w,u\in U,w\in W\}$ is a subspace of V.
3. If W is a vector space over an infinite field, then prove that V can not be expressed as the set theoretic union of a finite number of proper subspaces.
4. If V is a finite dimensional vector space, then show that any set of linearly independent vectors in V forms a basis of V.
5. If V is a finite dimensional vector space and W is a subspace of W such that $\dim V=\dim W$ then show that $V=W$.
6. Show that no linearly independent set in a vector space contains the zero vector. Also show that any subset of linearly independent set is linearly independent.
7. Show that the kernel of a linear map is a subspace.
8. Let $f:V\rightarrow V$ be a linear map defined on a vector space V and let $S=\{x\in V:\ f(x)=x\}$ show that S is a subspace of V.
9. If f is an isomorphism of V onto W then prove that f maps a basis of V onto a basis of W.
10. If V is a finite dimensional vector space and f is an isomorphism of V into itself, then f must be onto.
11. If V is finite dimensional and f is a linear map of V onto itself, then show that f must be 1–1 and hence is an isomorphism.

12. If V is finite dimensional and f is a linear map of V into itself which is not onto, then show that there exists $x \neq 0$ in V such that $f(x)=0$.
13. A linear map on a vector space V is 1–1 if and only if $f(x)=0$ implies $x=0$.
14. A linear functional which is not the zero functional is onto the field.
15. Given any two distinct vectors x and y in a vector space, show that there exists a linear functional f with $f(x) \neq f(y)$.
16. Given any two linearly independent vectors x and $y$in a vector space, show that there exists a linear functional f with $f(x)=f(y) \neq 0$.
17. Suppose f and g are linear functionals on a vector space and $g(x)=0$ for each x such that $f(x) = 0$. Show that g is a scalar multiple of f.
18. If A and B are submodules of module M then show that
    (a) $A \cap B$ is a submodule of M.
    (b) $A+B=\{a+b:\ a \in A,\ b \in B\}$ is a submodule of M.
19. If T is a homomorphism of a module M onto another module N with kernel $T = A$ then prove that N is isomorphic to M/A.
20. If T is a homomorphism of a module M into another module N then show that T is an isomorphism if and only if kernel $T=\{x \in M:\ T(x)=0\}=\{0\}$.
21. If T is a homomorphism of a module M into another module N then show that T is an isomorphism if and only if kernel $T=\{x \in M:\ T(x)=0\}$ and image $T=\{T(x):\ x \in M\}$ are respectively submodules of M and N.
22. A real square matrix $A=(a_{ij})$ of order n is said to be symmetric if:

    $A^T = A$ i.e. $a_{ij} = a_{ji}$.

    And skew-symmetric if:

    $A^T = -A$ i.e. $a_{ij} = -a_{ji}, i,\ j = 1, 2, 3,..., n$.

    Show that any real square matrix $A=(a_{ij})$ can be written as the sum of a symmetric matrix R and a skew-symmetric matrix S where $R=\frac{1}{2}\left(A+A^T\right)$ and $S=\frac{1}{2}\left(A-A^T\right)$.
23. Let A be a square matrix of order n. Let $\det A$ and $adj\ \ A$ respectively denote the determinant and the adjoint of A. Show that A^{-1} exists and is unique if and only if $\det\ \ A \neq 0$ and $A^{-1} = \dfrac{adj\ \ A}{\det\ \ A}$.

24. A square matrix is nonsingular if its determinant is nonzero. Show whether each of the following matrices is nonsingular and find inverse of nonsingular ones:

$$\begin{pmatrix} 3 & 5 \\ 2 & 4 \end{pmatrix}, \quad \begin{pmatrix} 4 & 7 \\ 1 & 6 \end{pmatrix}, \begin{pmatrix} 2 & 0 & 1 \\ 3 & 1 & 2 \\ 4 & 0 & 3 \end{pmatrix}.$$

If $A = \begin{pmatrix} \cos\theta & -\sin\theta \\ \sin\theta & \cos\theta \end{pmatrix}$ then show that $A^n = \begin{pmatrix} \cos n\theta & -\sin\ n\theta \\ \sin\ n\theta & \cos\ n\theta \end{pmatrix}$.

CHAPTER 7

Linear Transformation

7.1 ALGEBRA OF LINEAR TRANSFORMATION

In chapter 6 (Section 6.2) we have discussed about linear transformations. This chapter deals with algebra of linear transformations (Section 7.1) and eigen-values and eigen-vectors (Section 7.2).

Before we proceed further we first quote the definition of an algebra.

An algebra X over a field F is a vector space over F with a mapping $(x, y) \rightarrow xy$ of $X \times X$ into X called multiplication making X into a ring where scalar multiplication and multiplication are related in the following manner,

$$\lambda(xy) = (\lambda x)\, y = x(\lambda y) \text{ for } x, y \in X \text{ and } \lambda \in F.$$

The reader may try to prove that Hom (V), the set of all linear transformations of V to V is an algebra.

Theorem 7.1

If A is an algebra with unit element over F, then A is isomorphism into a subalgebra of Hom (V) for some vector space V over F.

Proof: A is an algebra over F.
$\Rightarrow A$ is a vector space over F.

We shall take $V = A$ to prove the theorem.

If $a \in A$ define $T_a: A \rightarrow A$ by

$$T_a v = va \, \forall v \in A$$

T_a is a linear transformation on $V = A$ i.e. $T_a \in$ Hom (A).

$$T_a(v_1 + v_2) = (v_1 + v_2)a = v_1 a + v_2 a = T_a v_1 + T_a v_2$$
$$T_a(\alpha v) = (\alpha v)a = \alpha(v_a) = \alpha T_a \,.$$

Consider the map: $\Psi: A \rightarrow \text{Hom}(A)$ defined by

$$\Psi(\mathrm{a}) = \mathrm{T_a} \forall a \in A$$

ψ is an isomorphism of A into Hom(A).

$$\Psi_{(\alpha a+\beta b)} = \mathrm{T}_{(\alpha a+\beta b)}$$

$$\mathrm{T}_{(\alpha a+\beta b)} = v(\alpha a + \beta b) = \alpha(va) + \beta(vb)$$

$$= \alpha T_a(v) + \beta T_b(\mathrm{v}) = (\alpha T_a + \beta T_b)(\mathrm{v})$$

Hence, $T_{(\alpha a+\beta b)} = \alpha T_a + \beta T_b$.

$\Rightarrow \Psi$ is a homomorphism (vector space homomorphism).

$$T_{ab} = T_b T_a$$

$$(T_b T_a)(v) = T_b(T_a v)$$

$$= T_b(v)$$

$$= vab$$

$$T_{ab} = v(ab).$$

ψ is a ring homomorphism.
$\Rightarrow \psi$ is a homomorphism of A into Hom(A).

It remains to show that ψ is $1 - 1$ i.e. Ker $\psi = \{0\}$.

Lat $a \in Ker\ \Psi \Rightarrow \Psi\mathrm{a} = 0 \Rightarrow \mathrm{T_a} = 0$

$\Rightarrow \mathrm{T_a V} = 0\ \forall\ \mathrm{v} \in V \Rightarrow \mathrm{T_a e} = 0$

So $0 = T_a e = ea = a\ \forall\ \mathrm{a} \in A$

$\Rightarrow \mathrm{a} = 0 \Rightarrow Ker\ \Psi = \{0\}$.

Theorem 7.2

Let A be an algebra with unit element over F, A is of dimension m over F. Then every element of A satisfies a nontrivial polynomial in $F[x]$ of degree atmost m.

Proof: Let e be the unit element of A. If a $\in A$, consider the $m + 1$ elements $e, a, a^2, \ldots, a^m$ all in A. Dim $A = m \Rightarrow e, a, a^2, \ldots\ l.d. \Rightarrow \exists$ elements $\alpha_0, \alpha_1, \ldots, \alpha_m \in F$, not all 0, such that,

$$\alpha_0 e + \alpha_1 a + \ldots + \alpha_m a^m = 0$$

$\Rightarrow$ a satisfies the nontrivial polynomial,

$$q(x) = \alpha_0 + \alpha_1 x + \alpha_2 x^2 + \ldots + \alpha_m x^m$$

of degree atmost m.

Theorem 7.3

Let V be an n-dimensional vector space over F. Given any element $T \in Hom(V), \exists\ a$ nontrivial polynomial $q(x) \in F(x)$ of degree atmost n^2 such that $q(T) = 0$.

Proof: $Hom(V)$ is of dimension n^2 over F and $Hom(V)$ is an algebra over F.

Hence T satisfies a nontrivial polynomial $q(x) \in F[x]$ of degree at most n^2.

Definition

1. $T \in Hom(V)$ is right-invertible if $\exists\ S \in Hom(V)$ such that $TS = 1$.
2. Left invertible if $\exists\ U \in Hom(V)$ such that $UT = 1$.
3. T is both right-and left-invertible if:

   $TS = UT = 1$, then

   $S = U$ and S is unique.
4. T is invertible or regular if it is both left and right invertible, i.e. if $\exists\ U \in Hom(V)$ such that

   $TS = UT = 1$ and $S = T^1$.
5. If T is not regular it is called singular.

Example 7.1

Right-invertible but not-invertible.

$$F = R, V = F[x]$$

In V define S and T by,

$$S(q(x)) = \frac{d}{dx}q(x)$$

$$T(q(x)) = \int_1^x q(x)dx$$

$$ST = 1, TS \neq 1.$$

Examples of Algebra

Example 7.2

Algebra of Quaternions over a field F.

Quaternion algebra is one with elements,

$$1, i, j, k$$

as a basis and

$$i^2 = -1, j^2 = -1, k^2 = -1,$$
$$ij = k, ji = -k, jk = -i,$$
$$kj = i, ki = j, ik = -j$$

This is an algebra of degree 4 with unit element 1.

Any element of this algebra looks like,

$$a + bi + cj + dk, a, b, c, d \in F.$$

$$\begin{aligned} XY &= (a_0 + a_1 i + a_2 j + a_3 k)(b_0 + b_1 i + b_2 j + b_3 k) \\ &= (a_0 b_0 - a_1 b_1 - a_2 b_2 - a_3 b_3) \\ &\quad +(a_0 b_1 + a_1 b_0 + a_2 b_3 - a_3 b_2)i \\ &\quad +(a_0 b_2 + a_2 b_0 + a_3 b_1 - a_1 b_3)j \\ &\quad +(a_0 b_3 + a_3 b_0 + a_1 b_2 - a_2 b_1)k \end{aligned}$$

Example 7.3

$C[0,1]$ = the set of all continuous functions real (complex) defined on [0,1] is an algebra with point wise multiplication of functions.

Example 7.4

A = the set of all complex functions analytic within unit disc, continuous within and on the disc is an algebra.

Example 7.5

$$l_1 = \left\{x: \sum_{-\infty}^{\infty} |x_n| < \infty\right\}$$

$$(x \times y)_n = \sum_{k=-\infty}^{\infty} x_{n-k} y_k.$$

Example 7.6

$$L_1(-\infty, \infty)$$

$$(f \times g)(x) = \int_{-\infty}^{\infty} f(x-y) g(y) dy$$

Theorem 7.4

If V is finite dimensional over F, then $T \in Hom(V)$ is invertible iff the constant term of the minimal polynomial for T is not zero.

Proof: Let $p(x) = \alpha_0 + \alpha_1 x + \cdots + \alpha_k x^k$, $\alpha_k \neq 0$, be the minimal polynomial for T over F.

If $\alpha_k \neq 0$, since

$$0 = \alpha_k T^k + \alpha_{k-1} T^{k-1} + \cdots + \alpha_1 T^1 + \alpha_0$$

we obtain,

$$\alpha_k T^k + \alpha_{k-1} T^{k-1} + \cdots + \alpha_1 T^1 = -\alpha_0$$

or $$\frac{-1}{\alpha_0}\left(\alpha_k T^k + \alpha_{k-1} T^{k-1} + \cdots + \alpha_1 T^1\right) = 1$$

or $$1 = \left(\frac{-1}{\alpha_0}\left(\alpha_k T^{k-1} + \cdots + \alpha_1\right)\right) T$$

Therefore,

$$S = \frac{-1}{\alpha_0}\left(\alpha_k T^{k-1} + \cdots + \alpha_1\right)$$

acts as an inverse for T,when T is invertible,

Suppose T is invertible, yet $\alpha_0 = 0$. Thus

$$\begin{aligned} 0 &= \alpha_1 T + \alpha_2 T^2 + \cdots + \alpha_k T^k \\ &= \left(\alpha_1 + \alpha_2 T + \cdots + \alpha_k T^{k-1}\right) T \end{aligned}$$

Multiplying this by T^{-1} from the right,

$$\alpha_1 + \alpha_2 T + \cdots + \alpha_k T^{k-1} = 0$$

Hence T satisfies the polynomial.

Since the degree of $q(x)$ is less than that of $p(x)$, this is impossible. Hence $\alpha_o \neq 0$.

Corollary 1

V is finite dimensional over $F, T \in Hom(V)$ is invertible $\Rightarrow T^{-1}$ is a polynomial expression in T over F.

Proof: By above theorem, $\alpha_0 + \alpha_1 x + \cdots + \alpha_k x^k = 0, \alpha_k \neq 0$.

$$T^{-1} = \frac{-1}{\alpha_0}\left(\alpha_k T^{k-1} + \cdots + \alpha_1\right)$$

Corollary 2

V is finite dimensional over $F, T \in Hom(V)$ is singular $\Rightarrow \exists\, S \neq 0$ in $Hom(V)$ such that $ST = TS = 0$.

Proof: ***T*** not regular $\Rightarrow$ constant term of minimal polynomial is 0, i.e.,

$$p(x) = \alpha_0 + \alpha_1 x + \cdots + \alpha_k x_k.$$

$\Rightarrow 0 = \alpha_1 T + \alpha_2 T^2 + \cdots + \alpha_k T^k$

Put $S = \alpha_1 + \alpha_2 T + \cdots + \alpha_k T^{k-1}$

S is of lower degree than $p(x) \Rightarrow S \neq 0$

$$ST = TS = 0.$$

Corollary 3

V is finite dimensional, $T \in Hom(V)$ is right-invertible $\Rightarrow T$ is invertible.

Proof: Let $TU = 1$. If T were singular, there would be an $S \neq 0$ such that $ST = 0$.

$$0 = (ST)U = S(TU) = S_1 = S \neq 0$$

$\Rightarrow$ Contradiction.

Theorem 7.5

V is finite dimensional over $F, T \in Hom(V)$ is singular, $\Rightarrow \exists\, v \neq 0$ in V such that $Tv = 0$.

Proof: By Corollary 2, T is singular $\Leftrightarrow S \neq 0$ in $Hom(V)$ such that $ST - TS = 0$. Since S 0 $S \neq 0, \exists\, v \in V$ such that $Sw \neq 0$.

Let $v = Sw$. Then

$$Tv = T(Sw) = (TS)w = 0(w) = 0.$$

Conversely let $Tv = 0$ with $v \neq 0 \Rightarrow T$ is not invertible.

Suppose to the contrary that T invertible

$\Rightarrow TT^{-1} = 1$. Then

$$v = 1(v) = (TT^{-1})v = T^{-1}\,(Tv) = T^{-1}\,(o) = 0$$

$\Rightarrow v = 0 \Rightarrow$ contradiction.

Theorem 7.6

V is finite dimensional over $F, T \in Hom(V)$ is regular $\Leftrightarrow T$ maps V onto V, that is $TV = V$.

Proof: Let T be regular. Then given $v \in V, v = T(T^{-1}v) \Rightarrow T$ is onto.

$\Leftarrow$: Assume that T is not regular.

We shall show that T is not onto. T is singular $\Rightarrow v_1 \neq 0$ in V such that $Tv_1 = 0$.

[We use the following result to complete the proof:

If V is finite dimensional over F and if $u_1, u_2, \dots, u_m \in V$ are 1.i., then we can find vectors $\{u_{m+1}, u_{m+2}, \dots, u_{m+r}\}$ in V such that $\{u_1, u_2, \dots, u_m, u_{m+1}, u_{m+2}, \dots, u_m$ in V is a basis of V].

By this result we can fill out, from v_1 to a basis $\{v_1, v_2, \dots, v_n\}$ of V. Then every element of TV is a linear combination of the elements.

$$w_1 = Tv_1, w_2 = Tv_2, \dots, w_n = Tv_n.$$

Since $w_1 = 0$, VT is spanned by the $(n-1)$ elements $w_2, \dots, w_n$; hence

$$dim\, VT \leq (n-1) < n = dim\, V.$$

$\Rightarrow VT$ is different from V.

$\Rightarrow T$ is not onto.

7.2 CHARACTERISTIC ROOTS (EIGEN-VALUES)

Let V be finite dimensional over F. Let $Hom(V)$ have a unit element 1 and $\lambda - T$ mean $\lambda 1 - T$.

Definition

If $T \in Hom(V), \lambda \in F$ is called a characteristic root(eigen value) of T if $\lambda - T$ is singular.

Theorem 7.7

$\lambda \in F$ is characteristic root if for some $v \neq 0, Tv = \lambda v$.

Proof: λ is a characteristic root $\Rightarrow \lambda - T$ is singular $\Rightarrow \exists\, v \neq 0$ such that $(\lambda - T)(v) = 0$

$\Rightarrow Tv = \lambda v.$

If $Tv = \lambda v$ for $v \neq 0$ in V, then $(\lambda - T)(v) = 0 \Rightarrow \lambda - T$ singular $\Rightarrow \lambda$ is a characteristic root.

Theorem 7.8

If $\lambda \in F$ is a characteristic root of $T \in Hom(V)$, then for any polynomial $q(x) \in F[x], q(\lambda)$ is a characteristic root of $q(T)$.

Proof: λ is a characteristic root of T.

$\Rightarrow \exists\, v \neq 0$ such that $Tv = \lambda v$

Now, $T^2 v = T(Tv) = T(\lambda v) = \lambda(Tv)$

$$= \lambda(\lambda v) = \lambda^2 v$$

Continuing in this way, we get

$$T^k v = \lambda^k v \text{ for all positive integers } k.$$

If $q(x) = \alpha_0 x^m + \alpha_1 x^{m-1} + \cdots + \alpha_m, \alpha_0 \in F$, then

$$q(T) = \alpha_0 T^m + \cdots + \alpha_m$$

$$\begin{aligned}\Rightarrow (q(T))v &= \alpha_0 T^m v + + \cdots + \alpha_m v \\ &= \alpha_0 \lambda^m v + + \cdots + \alpha_m v \\ &= (\alpha_0 \lambda^m + + \cdots + \alpha_m)v \\ &= q(\lambda)(v)\end{aligned}$$

$\Rightarrow (q(\lambda) - q(T))(v) = 0$

$\Rightarrow q(\lambda)$ is a characteristic root of $q(T)$.

Theorem 7.9

If $\lambda \in F$ is a characteristic root of $T \in Hom(V)$, then λ is a root of the minimal polynomial of T. T only has finite number of characteristic roots in F.

Proof: Let $p(x)$ be the minimal polynomial over F of T. $\Rightarrow p(T) = 0$. If λ is a characteristic root of T, then $\exists\, v \neq 0$ such that $Tv = \lambda v$,

$\Rightarrow (p(T))(v) = p(\lambda)v$.

But $p(T) = 0$.

Hence $p(\lambda)v = 0$.

But $p(T) = 0$. Hence $p(\lambda)v = 0$. But $\lambda \neq 0$, Hence $p(\lambda) = 0$.

$\Rightarrow \lambda$ is root of $p(x)$. $p(x)$ has only finite roots, $\deg(p(x) \leq n^2, n = \dim V)$, $p(x)$ has atmost n^2 roots.

$\Rightarrow$ There can be only finite number of characteristic roots of T in F.

Theorem 7.10

If $T, S \in Hom(V)$ and S is regular, then T and STS^{-1} have the same minimal polynomial.

Proof: $(STS^{-1})^2 = STS^{-1}STS^{-1} = ST^2S^{-1}$

$$(STS^{-1})^3 = ST^3S^{-1}$$

$$\vdots$$

$$(STS^{-1})^i = ST^iS^{-1}$$

For any $q(x) \in F[x], q(STS^{-1}) = Sq(T)S^{-1}$.

If $q(T) = 0$, then $q(STS^{-1}) = 0$. Thus if $q(x)$ is the minimal polynomial for T, then $q(x)$ is also the minimal polynomial for STS^{-1}.

Definition

$0 \neq v$ is called the characteristic vector corresponding characteristic root λ, if $Tv = \lambda . v$.

Theorem 7.11

If $\lambda_1, \lambda_2, \ldots, \lambda_k$ are distinct characteristic roots of T and if $v_1, v_2, , \ldots, v_k$ are characteristic vectors of T belonging to $\lambda_1, \lambda_2, \ldots, \lambda_k$ respectively, then $v_1, v_2, , \ldots, v_k$ are ℓ.i.

Proof: Let $k > 1$.
If $v_1, v_2, \ldots, v_k$ are ℓ.d., then $\exists$ a relation $\alpha_1 v_1 + \cdots + \alpha_k v_k = 0, \alpha_1, \alpha_2, \ldots,$ $\alpha_k \in F$, not all 0.

In all such relations, $\exists$ one having as few nonzero coefficients as possible. By suitably renumbering the vectors, we can assume this shortest relation to be,

$$\beta_1 v_1 + \ldots + \beta_j . v_j = 0, \beta_1 \neq 0, \ldots, \beta_j \neq 0 \quad \ldots (1)$$

$$Tv_i = \lambda_i v_i$$

$$\beta_1 Tv_1 + \ldots + \beta_j Tv_j = 0$$

Or $$\lambda_1 \beta_1 v_1 + \ldots + \lambda_j \beta_j . v_j = 0 \quad \ldots (2)$$

Multiply (1) by λ_1 and subtract from (2):

$$(\lambda_2 - \lambda_1)\beta_2 v_2 + \cdots + (\lambda_j - \lambda_1)\beta_j v_j = 0.$$

Now $\lambda_i - \lambda_1 \neq 0$ for $i > 1$, also $\beta_i \neq 0$.

$\Rightarrow (\lambda_i - \lambda_1)\beta_i = 0.$

$\Rightarrow$ We have produced a shorter relation than that in (1) between $v_1, v_2, \dots, v_k$. This is a contradiction and hence the theorem follows.

Corollary

If $\dim V = n, T$ can have atmost n distinct characteristic roots.

Proof: Suppose T has more than n distinct characteristic roots. The corresponding characteristic vectors are $\ell.i.$ They are more than n in number. It is impossible since $\dim V = n,$ any set of $\ell.i.$ elements in V can have atmost n elements.

Corollary

Dim $V = n.$ If T has n distinct characteristic roots in F, then $\exists$ a basis of V over F which consists of characteristic vectors of T.

Remark

Every linear transformation ($\ell.t.$) on a vector space over any field F may not have all its characteristic roots in F. This depends on the field. For example take $F = R$. Then the minimal polynomial of

$$\begin{pmatrix} 0, 1 \\ -1, 0 \end{pmatrix}$$

over F is $x^2 + 1$. This has no roots in F. Hence we cannot assume that characteristic roots always lie in the field in question. However, things work in some extension of F; every root of the minimal polynomial of T over F is a characteristic root of T. But an important fact is that any polynomial with coefficients in C has all its roots in C. This is called the Fundamental Theorem of Algebra. One implication from this is that every ℓ.t. will have all its characteristic roots in C. Let V denote a finite-dimesional inner-product space over C and $T \in Hom(v)$.

Theorem 7.12

$(Tv, v) = 0\ \forall v \in V \Rightarrow T = 0.$

Proof: $(Tv, v) = 0\ \forall v \in V.$

$\Rightarrow$ Given $u, w, \in V,$

$$\begin{aligned} 0 &= (T(u+w), u+w) \\ &= (Tu + Tw, u+w) \\ &= (Tu, u) + (Tu, w) + (Tw, u) + (Tw, w) \\ &= (Tu, u) + (Tw, u) \end{aligned} \quad \ldots (1)$$

Equation (1) holds if w is replaced by iw. So,

$$0 = (Tu, iw) + (T(iw), u) = -i(Tu, w) + i(Tw, u)$$

So $(Tw, u) - (Tu, w) = 0$ ... (2)

Adding (1) and (2) we get, $(Tw, u) = 0\ \forall u \in V$
$\Rightarrow Tw = 0\ \forall\ w \in V \Rightarrow T = 0.$

Remark

If T is an ips over R, then the result is false.

Take $V = R^2 = \{(\alpha, \beta);\ \alpha, \beta \in R\}$

and $(\alpha, \beta) = \alpha\beta$

$$T(\alpha, \beta) = (-\beta, \alpha)$$

$$(Tv, v) = 0 \text{ yet } T \neq 0.$$

$$\begin{aligned} 0 = (Tv, v) &= \big(T(\alpha, \beta), (\alpha, \beta)\big) = (-\beta, \alpha).(\alpha, \beta) \\ &= -\alpha\beta + \alpha\beta = 0. \end{aligned}$$

Definition

T is called unitary iff $(Tu, Tv) = (u, v)\ \forall\ u, v\ \in V,$

$$||v|| = \sqrt{(v, v)} = \sqrt{(Tv, Tv)} = ||Tv||.$$

Unitary operator preserves length and the converse is also true. We have

Remark

$(Tv, Tv) = (v, v) \Rightarrow T$ is unitary.

Proof: $(T(u+v),\ T(u+v)) = (u+v, u+v)$

Or $(Tu+Tv,\ Tu+Tv) = \big((u+v),(u+v)\big)$

Or $(Tu,Tu)+(Tu,Tv)+(Tv,\ Tu)+(Tv,\ Tv)$

$$= (u,u)+(u+v)+\ (v+u)+(v,v)$$

Or $(Tu,\ Tv)+(Tv,\ Tu) = (u,v)+(v,u)$ ... (1)

Replace v by iv:

$$-(Tu,Tv)+(Tv,Tu) = -(u,v)+(v,u) \quad \ldots (2)$$

Add (1) and (2). Then we get $(Tu,Tv) = (u,v)$

Theorem 7.13

$T \in Hom(V)$ is unitary $\Leftrightarrow$ it takes an orthonormal basis of V into an orthonormal basis of V.

Proof: Suppose $\{v_1, v_2, \ldots v_n\}$ is an arthonormal bases of V.

$$\text{S}_\text{o}\ (v_i, v_j) = \begin{cases} 0\ i \neq j, \\ 1\ i = j \end{cases}$$

$\Rightarrow$ To show that

T unitary $\Rightarrow \{Tv_1, \ldots, Tv_n\}$ is an orthonormal basis of V. Since

$$(Tv_i, Tv_j)(v_i, v_j) = \begin{cases} 0\ i \neq j, \\ 1\ i = j \end{cases}$$

the result follows.

Conversely, T is such that

$\{v_1, \ldots, v_n\}, \{Tv_1, \ldots, Tv_n\}$ both are orthonormal bases of V. If $u, w \in V$, then

$$u = \sum_{i=1}^{n} \alpha_i v_i\,, w = \sum_{i=1}^{n} \beta_i v_i$$

$$(u,w) = \left(\sum_{i=1}^{n} \alpha_i v_i\,, \sum_{i=1}^{n} \beta_i v_i\right)$$

$= \sum_{i=1}^{n} \alpha_i \overline{\beta}_i$ since v_i's are orthonormal.

However,

$$Tu = \sum_{i=1}^{n} \alpha_i Tv_i\,, Tw = \sum_{i=1}^{n} \beta_i Tv_i$$

since Tv_i's are orthonormal

$$(Tu, Tw) = \sum_{i=1}^{n} \alpha_i \overline{\beta}_i = (u, w).$$

$\Rightarrow T$ is unitary and this proves the result .

Theorem 7.14

Let $T \in Hom(V)$. Given $v \in V, \exists\, w \in V$ such that $(Tu, v) = (u, w)\ \forall\, u \in V$. This w is

Uniquely determined by v and T.

Proof: It suffices to exhibit a $w \in V$ which works for all the elements of a basis of V.

Let $\{u_1, u_2, .., u_n\}$ be an orthonormal basis of V. Define

$$w = \sum_{i=1}^{n} \overline{(Tu_i, v)} ui.$$

Observe that,

$$(u_i, w) = (Tu_i, v)$$

Hence w has the desired property,

w is unique Suppose,

$$(Tu, v) = (u, w_1) = (u, w_2)$$

$$\Rightarrow (u, w_1 - w_2) = 0\ \forall\, u \in V$$
$$\Rightarrow (w_1 - w_2, w_1 - w_2) = 0$$
$$\Rightarrow w_1 - w_2 = 0 \Rightarrow w_1 = w_2.$$

Definition

Let $T \in Hom(V)$, Hermitian Adjoint of T, T^* is defined by $(Tu, v) = (u, T^*v)$.

Theorem 7.15

1. $(T^*)^* = T$,
2. $(S + T)^* = S^* + T^*$
3. $(\lambda S)^* = \overline{\lambda} S^*$
4. $(ST)^* = T^* S^*$

T^* is a linear transformation of V.

$$\begin{aligned}\big(u, T^*(v+w)\big) &= (Tu, u+w) = (Tu, v) + (Tu, w)\\ &= (u, T^*v) + (u, T^*w)\\ &= (u, T^*v + T^*w)\end{aligned}$$

$$\Rightarrow \big(T^*(v+w)\big) = (T^*v + T^*w)$$

$$\begin{aligned}\big(u, T^*(\lambda v)\big) &= (Tu, \lambda v) = \bar{\lambda}(Tu, v)\\ &= \bar{\lambda}(u, T^*v) = (u, \lambda(T^*v))\end{aligned}$$

$$\Rightarrow T^*(\lambda v) = \lambda(T^*v).$$

$$(T^*)^* = T$$

$$(u, (T^*)^*v) = (T^*u, v) = \overline{(v, T^*u)} = \overline{(Tv, u)} = (Tv, u)$$

$$\Rightarrow (T^*)^*v = Tv \Rightarrow (T^*)^* = T.$$

$$(ST)^* = T^*S^*.$$

$$\begin{aligned}(\mathrm{u}, (ST)^*v) &= (STu, v) = (Tu, S^*v)\\ &= (u, T^*S^*v)\end{aligned}$$

$$\Rightarrow (ST)^*v = T^*S^*v \Rightarrow (ST)^* = T^*S^*.$$

Theorem 7.16

T is unitary $\Leftrightarrow T^*T = I$.

Proof: T is unitary $\Rightarrow \forall u, v \in V$

$(u, T^*Tv) = (Tu, Tv) = (u, v) \Rightarrow T^*T = I.$

$T^*T = I \Rightarrow (u, v) = (u, T^*Tv) = (Tu, Tv)$

$\Rightarrow T$ is unitary.

Remark

Unitary transformation is nonsingular.

Its inverse is its Hermitian adjoint.

$$T^*T = I \Rightarrow TT^* = I$$

Definition

T is self-adjoint or Hermitian if $T^* = T$.

Skew-Hermitian if $T^* = -T$.

Fact. Any linear transformation S can be expressed as

$$S = A + Bi$$

where A and B are Hermitian.

Proof: Given $S \in Hom(V)$,

$$S = \frac{S + S^*}{2} + i\frac{S - S^*}{2i}$$

And $\frac{S+S^*}{2}, \frac{S-S^*}{2i}$ are Hermitian.

$S = A + Bi, A, B$ are Hermitian.

Theorem 7.17

$T \in Hom(V)$ Hermitian ⇒ all its characteristic roots are real.

Proof: Let λ be any characteristic root of T.

$\Rightarrow \exists v \neq 0$ such $Tv = \lambda v$. Now

$$\lambda(v, v) = (\lambda v, v) = (Tv, v) = (v, T^* v)$$
$$= (v, Tv) = (v, \lambda v) = \bar{\lambda}(v, v)$$

Since $(v, v) \neq 0, \lambda = \bar{\lambda}$
$\Rightarrow \lambda$ is real.

Theorem 7.18

T Hermitian $\Leftrightarrow (Tu, u)$ is real.

Proof: T Hermitian

$\Rightarrow (Tu, u) = (u, T^* u) = (u, Tu) = \overline{(Tu, u)}$

$\Rightarrow (Tu, u)$ is real.

Let (Tu, u) be real

$\Rightarrow \langle Tu, u\rangle = \langle\overline{Tu, u}\rangle$

$= \langle\overline{u, T^* u}\rangle$

$= \langle T^* u, u\rangle$

$\Rightarrow \langle T - T^*\rangle u, u >= 0$

$\Rightarrow T - T^* = 0$

Since $\langle Tv, v\rangle = 0 \Rightarrow T = 0$.

Theorem 7.19

T_1, T_2 Hermitian $\Longrightarrow$ T_1T_2 Hermitian iff $T_1T_2 = T_2T_1$.

Proof: $(T_1T_2)^* = T_2^*T_1^* = T_2T_1$.

Theorem 7.20

$T \in Hom(V)$ and $T^*Tv = 0 \Rightarrow TV = 0$.

Proof: $0 =< T^*Tv, v > =< Tv, (T^*)^*v > = < Tv, Tv > \Rightarrow Tv = 0$

Definition

$T \in Hom(V)$ is called normal iff $TT^* = T^*T$.

Remark

T Hermitian $\Rightarrow$ T normal
T unitary $\Rightarrow$ T normal

Theorem 7.21

N is normal and $Nv = 0\ \forall v \in V \Rightarrow N^*v = 0$.

Proof: $< N^*v, N^*v > = < NN^*v, v > = < N^*Nv, v >$

But $v = 0$, hence $N^*Nv = 0$

$\Rightarrow < N^*v, N^*v > = 0 \Rightarrow N^*v = 0$

Corollary

If N is normal and λ is a characteristic root of N i.e. $Nv = \lambda v$, then $N^*v = \bar{\lambda}v$.

Proof: N is normal $\Rightarrow NN^* = N^*N$

$\Rightarrow (N - \lambda)(N - \lambda)^* = (N - \lambda)(N^* - \bar{\lambda})$
$= NN^* - \lambda N^* - \bar{\lambda}N + \lambda\bar{\lambda}$
$= NN^* - \lambda N^* - \bar{\lambda}N + \lambda\bar{\lambda}$

$= N^*(N - \lambda) - \overline{\lambda}(N - \lambda)$
$= (N^* - \overline{\lambda})$
$= (N - \lambda)^*(N - \lambda)$

$\Rightarrow N - \lambda$ is normal

Now $Nv = \lambda v$

$\Rightarrow (N - \lambda)(v) = 0$
$\Rightarrow (N - \lambda)^*(v) = 0$
$\Rightarrow (N - \overline{\lambda})(v) = 0$
$\Rightarrow N^*v = \overline{\lambda}v$.

Corollary

If T is unitary and λ is a characteristic root of T, then $|\lambda| = 1$.

Proof: T is unitary $\Rightarrow T$ is normal. Let λ be a characteristic root of T and $Tv = \lambda v, v \neq 0$

$\Rightarrow T^*\ v = \overline{\lambda}v$. Thus

$$v = T^*Tv = T^*(\lambda v) = \lambda T^*v = \lambda\overline{\lambda}v.$$

Thus $\lambda\overline{\lambda} = 1 \ \Rightarrow\ |\lambda| = 1$.

Theorem 7.22

Let N be normal. Eigen vectors corresponding to different eigen –values are orthogonal.

Proof: Let N be normal. Let $\lambda \neq \mu, Nv = \lambda v, Nw = \mu w$. Then we have to show that $(v, w) = 0$.

Since $Nv = \lambda v$,

$$< Nv, w > = < \lambda v, w > = \ \lambda(v, w)$$

Since $Nw = \mu w$ we have

$N^*w = \overline{\mu}w$. Hence

$<Nv, w> = <v, N^*w> = <v, \overline{\mu}\, w> = \mu <v, w>$.

Hence, $< v, w > = 0$.

Theorem 7.23

If T_1, T_2 are normal operators such that each commutes with the adjoint of the other, then $T_1 + T_2, T_1, T_2$ are normal.

Proof: We have

$$(T_1 + T_2)(T_1 + T_2)^* = (T_1 + T_2)(T_1^* + T_2^*)$$
$$= T_1T_1^* + T_1T_2^* + T_2T_1^* + T_2^*T_2^*$$
$$(T_1 + T_2)^*(T_1 + T_2) = (T_1^* + T_2^*)(T_1 + T_2)$$
$$= T_1^* T_1 + T_1^* T_2 + T_2^* T_1 + T_2^* T_2$$

$\Rightarrow T_1 + T_2$ is normal.

$$T_1T_2(T_1T_2)^* = T_1T_2T_2^*T_1^* = T_1T_2^*T_2T_1^*$$
$$= T_2^*T_1T_1^*T_2 = T_2^*T_1^*T_1T_2$$
$$= (T_1T_2)^*T_1T_2 .$$

$\Rightarrow T_1T_2$ is normal.

Definition

T is a non-negative (positive semidefinite) operator if $< Tu, u > \geq 0 \ \forall u, T$ is positive (positive definite) if $< Tu, u >> 0$ for $u \neq 0$.

A Hermitian operator T is non-negative (positive) iff all its characteristic roots are non-negative (positive).

Proof: $\Rightarrow T$ is non negative.

If λ characteristic root, $Tv = \lambda v$ for some $v \neq 0$.

$$0 \leq < Tv, v > - < \lambda v, v > = \lambda < v, v >$$
$$< v, v > \geq 0 \Rightarrow \lambda \geq 0.$$

$\Leftarrow$: Let T be Hermitian with non negative characteristic roots. We can find an orthonormality basis consisting of characteristic vectors of T.

For each $v_i, Tv_i = \lambda_i v_i$ where $\lambda \geq 0$.

Given $v \in V, v = \sum \alpha_i v_i$.

Hence $Tv = T(\sum \alpha_i v_i) = \sum \alpha_i \lambda_i v_i$

But $< Tv, v > = < \sum \alpha_i \lambda_i v_i, \sum \alpha_i v_i >$

$= \sum \lambda_i \alpha_i \overline{\alpha_i}$ by orthonormality of v_i

Since $\lambda_i \geq 0, \alpha_i \overline{\alpha_i} \geq 0$ where

$$< Tv, v > \geq 0.$$

CHAPTER 8

Boolean Algebras

In this chapter we discuss the concepts of Lattices and Boolean Algebras. Before doing this, we first discuss the concept of order relations.

8.1 PARTIAL ORDER RELATION

Let X be any nonempty set. A partial order relation (por) in X is a relation $\leq \subset X \times X$ such that for all $x, y, z \in X$

We have

1. $x \leq x$ (reflexive)
2. $x \leq y$ and $y \leq x \Rightarrow x = y$ (antisymmetric);
3. $x \leq y$ and $y \leq z \Rightarrow x \leq z$ (transitive)

We sometimes write $x \leq y$ in the equivalent form $y \geq x$. A non empty set X with a partial order relation defined on it, is called a partially ordered set (poset). We now give three important and simple examples of partially ordered sets.

8.1.1 *Example*

(a) Let X be the set of all positive integers, and $x \leq y$ means that x divides y.

(b) Let X be the set of all real numbers, and $x \leq y$ means that x is less than or equal to y in usual sense.

(c) Let X be the family of all subsets of universal set U and let $A \leq B$ mean $A \subset B$.

Two elements x and y in a poset are called comparable if either $x \leq y$ or $y \leq x$. Note that there may be pairs of elements in a poset which are not comparable. In Example 8.1.1(a), for example, the integers 3 and 7 are not comparable because neither divides the other; and in Example 8.1.1(c) if U has more than one element, there are always two subsets of U neither of which is a subset of the other.

Some partial order relations possess a fourth property in addition to the three required by the definition.

Any two elements are comparable.

A por with this property is called a total (or linear) order relation and a set with a total (or linear) order relation defined on it is called a totally ordered set or a linearly ordered set or a chain.

8.2 LATTICES AND BOOLEAN ALGEBRAS

Let $(X, \leq)$ be a poset and let $A \subset X$. An element $x \in X$ is said to be an upper bound for A if $a \leq x$ for every $a \in A$. x is said to be the least upper bound $(l.u.b)$ or supremum (if it exists) for A if x is an upper bound for A and if y is any other upper bound for A, then $x \leq y$. Similarly we define a lower bound and the greatest lower bound $(g.l.b)$ or infimum for a subset of a poset.

A poset $(X, \leq)$ is said to be a lattice if every two element subset of X (i.e., every pair of elements of X) has both the supremum and infimum.

If we define the infimum of x and y to be $x \wedge y$ and the supremum of x and y to be $x \vee y$, then these elements are uniquely detemined by x and y. The operations $\wedge$ and $\vee$ have the following properties.

Idempotent laws: $x \vee x = x, x \wedge x = x$
Commutative laws: $x \vee y = y \vee x, x \wedge y = y \wedge x$
Associative laws: $x \vee (y \vee z) = (x \vee y) \vee z, (x \wedge y) \wedge z = (x \wedge y) \wedge z$
Absorption laws: $x \vee (x \wedge y) = x, x \wedge (x \vee y) = x$

Each of these laws consists of two identities. In each case we prove only one identity. Observe that $x \leq x \vee x$. By reflexivity, $x \leq x$ and so $x \vee x \leq x$. Thus $x \wedge x = x$.

Now $y \leq x \vee y$ and $x \leq x \vee y$. Hence $y \vee x \leq x \vee y$.

Similarly $x \leq y \vee x$ and $y \leq y \vee x$. Therefore, $x \vee y \leq y \vee x$.

Thus by antisymmetry $x \vee y = y \vee x$.

Let $a = x \vee (y \vee z)$ and $b = (x \vee y) \vee z$. Then $x \leq a, y \vee z \leq a$, i.e., $y \leq a$ and $z \leq a$. Now $x \leq a$ and $y \leq a$ imply $x \vee y \leq a$.

Again $x \vee y \leq a$ and $z \leq a$ imply $b = (x \vee y) \vee z \leq a$. Similarly, it can be shown that $a = x \vee (y \vee z) \leq b$; and hence by antisymmetry $a = b$.

We have $x \leq \vee (y \wedge z)$. Also $x \wedge y \leq x$ and hence $x \vee (x \wedge y) \leq x$. Thus by antisymmetry.

$$x = x \vee (x \wedge y).$$

We now show that the above properties actually characterise lattices. We prove the following:

Let X be a nonempty set over which two operations $\wedge$ and $\vee$ are defined and let these operations satisfy the above four properties. Then a partial order relation can be defined on X such that X becomes a lattice in which $x \vee y$ and $x \wedge y$ are the supremum and infimum of x and y.

First, observe that $x \wedge y = x$ and $x \vee y = y$ are equivalent, for if $x \wedge y = x$ and $x \vee y = (x \wedge y) \vee y = (y \wedge x) \vee y = y$ and similarly, $x \vee y = y$ implies $x \wedge y = x$. We now define a relation $' \leq '$ on X as follows:

$x \leq y$ means that either $x \wedge y = x$ or $x \vee y = y$. Since $x \wedge x = x$ we have $x \leq x$ for each $x \in X$, so that the relation, $' \leq '$ is reflexive.

If $x \leq y$ and $y \leq x$, so that $x \wedge y = x$ and $y \wedge x = x$, then $x = x \wedge y = y \wedge x = y$. So that relation $' \leq '$ is anti-symmetric. Now if $x \leq y, y \leq z$, so that $x \wedge y = x$ and $y \wedge z = y$, then

$x \wedge z = (x \wedge y) \wedge z = x \wedge (x \wedge y) = x \wedge (y \wedge z) = x \wedge y = x$, so $x \leq x$ and hence the relation $' \leq '$ is transitive.

We now show that $x \wedge y$ is the infimim of x and y. Since $(x \wedge y) \vee x = x$ and $(x \wedge y) \vee y = (y \wedge x) \vee y = y$, we have $x \wedge y \leq x$ and $x \wedge y \leq y$.

If $z \leq x$ and $z \leq y$, so that $z \wedge x = z$ and $z \wedge y = z$, then $z \wedge (x \wedge y) = (z \wedge x) \wedge y = z \wedge y = z$, so $z \leq x \wedge y$. Thus $x \wedge y$ is the glb or infimum of x and y. In a similar vein, it can be shown that $x \vee y$ is the lub or supremum of x and y.

A lattice is said to be distributive if the following distributive laws holds,

$$x \vee (y \wedge z) = (x \vee y) \wedge (x \vee z)$$

$$x \wedge (y \vee z) = (x \wedge y) \vee (x \wedge z)$$

A Lattice is said to be complemented if it contains distinct elements 0 and 1 such that $0 \leq x \leq 1$

for every x and if each element x has a complement x' with the property that,

$$x \vee x' = 1, x \wedge x' = 0$$

A Boolean algebra is a complemented distributive lattice.

It is possible for an element of complemented lattice to have more than one complement. But in a Boolean algebra each element has a unique complement. To prove this assume that x^* is also an element with the property that $x \wedge x^* = 0$ and $x \vee x^* = 1$.

Then

$$x = x \wedge 1 = x \wedge (x \vee x') = (x \wedge x) \vee (x \wedge x')$$
$$= 0 \vee (x \wedge x') = x \wedge x'$$

So $x^* \leq x'$, similarly it can be shown that,

$x' \leq x^*$ and so $x^* = x'$.

Also it is clear that the complement of x' (complement of complement of x) is x . i.e.,

$$x'' = x.$$

Furthermore, $0 \wedge 1 = 0$ and $0 \vee 1 = 1$, so we have

$0' = 1$ and $1' = 0$

In a Boolean algebra we have

$x \leq y$ implies $y' \leq x'$.

Now if $x \leq y$, then $x \wedge y' \leq y \wedge y' = 0$ and hence

$$y' = y' \wedge 1 = y' \wedge (x \vee x') = (y' \wedge x) \vee (y' \vee x')$$
$$= 0 \vee (y' \wedge x') = y' \wedge x' .$$

Therefore $y' \leq x'$ and this proves that result.

Finally we show that in a Boolean algebra the following identities hold:

$(x \wedge y)' = x' \vee y'$ and $(x \vee y)' = x' \wedge y'$.

We only prove the first one. Observe that if $x' \leq z$ and $y' \leq z$, so that $z' \leq x$ and $z' \leq y$.

Then $z' \leq x \wedge y$ or $(x \wedge y)' \leq z$. This shows that $(x \wedge y)' \leq$ any upper bound of x' and y', so $(x \wedge y)' \leq x' \vee y'$. Therefore it remains to show that $x' \vee y' \leq (x \wedge y)'$. Now since $x \wedge y \leq x$ and $x \wedge y \leq y$ we have $x' \leq (x \wedge y)'$ and $y' \leq (x \wedge y)'$ and hence the result holds.

We may recall that a Boolean algebra is a complemented distributive lattice. To be more explicit a Boolean algebra $B = (B, \wedge, \vee, /, 0, 1)$ is a set B with two binary operations ($\wedge$, $\vee$), two universal bounds (0, 1) and one unary operation such that for all $x, y, z \in B$.

1. $x \wedge x = x, x \vee x = x$ (Idempotence)
2. $x \wedge y = y \wedge x, x \vee y = y \vee x$ (Commutativity)
3. $x \vee (y \vee z) = (x \vee y) \vee z, (x \wedge y) \wedge z = (x \wedge y) \wedge z$ (Associativity)
4. $x \vee (x \wedge y) = x, x \wedge (x \vee y) = x$ (Absorption)

5. $x \wedge [x \vee (x \wedge z)] = (x \wedge y) \vee (x \wedge z)$,
   $x \vee [x \wedge (x \vee z)] = (x \vee y) \wedge (x \vee z)$
6. $x \vee (y \wedge z) = (x \vee y) \wedge (x \vee z), x \wedge (y \vee z) = (x \vee y) \wedge (x \vee z)$ (distributivity)
7. $x \wedge 0 = 0, x \vee 0 = x$
   $x \wedge 1 = 1, x \vee 1 = 1$ (laws of the bounds)
8. $x \wedge x' = 0, x \vee x' = 1$ (laws of the complementation)
9. $(x')' = x$ (Laws of involution)
10. $(x \wedge y)' = x' \vee y', (x \vee y)' = x' \wedge y'$ (De Morgan Laws)

8.3 DUALITY PRINCIPLE

In united states all passenger cars have the following traffic regulations:

Cars always travel on the right of the street. On a multilane divided highway, left lanes are passing lanes. Cars may make left turns at a red light.

In England all passenger cars have the following traffic regulations:

Cars always travel on the left of the street. On a multilane divided highway, right lanes are passing lanes. Cars may make right turns at a red light.

This is an example of duality principle. Let us take a mathematical example. Let $(A, \leq)$ be a poset. If we define $(a \frac{\leq}{R} b)$ iff $b \leq a$, then $\left(A, \frac{\leq}{R}\right)$ is another poset. Further if $(A, \leq)$ is a lattice, then $\left(A, \frac{\leq}{R}\right)$ is also a lattice. The meet "$\wedge$" and join "$\vee$" operations with respect to $(A, \leq)$ will respectively be the join and meet operations with respect to $\left(A, \frac{\leq}{R}\right)$; in other words, the glb and lub for $(A, \leq)$ are respectively the lub and glb for $(A, \leq)$. This is a case of duality principle. Each of the laws stated above for lattices and Boolean algebras has two identities, one is the dual of the other.

8.4 BOOLEAN EXPRESSION

An element $x \in B$ is called a Boolean variable. If a Boolean variable takes two distinct values 0 and 1 (which denotes in particular the states "off" and "on" respectively of a switch or "No" and "Yes" respectively of a logical expression then it is called a binary Boolean variable. In applications usually binary Boolean variables are used and for convenience we call them simple Boolean variables.

A Boolean expression is an expression containing a finite number for symbols each representing a variable element and the operations $\wedge$, $\vee$, $'$. For example, $((x' \wedge y) \vee z) \vee (x \wedge y')$ is Boolean expression.

If $x_1, x_2, \ldots, x_n$ are Boolean variables, then we write $(x_1, x_2, \ldots, x_n)$ to denote a Boolean expression.

Let $(B, \vee, \wedge, ')$ be a Boolean algebra. To be more precise, a Boolean expression over this Boolean algebra is defined as follows:

(i) any element of A is a Boolean expression,
(ii) any variable name is a Boolean expression.
(iii) if e_1 and e_2 are Boolean expressions, then $e_1', e_1 \vee e_2, e_1 \wedge e_2$ are Boolean expressions.

Observe that if a Boolean expression and the values of the variables are given, we can evaluate the expression by substituting the values of the variables in the expression. For example, consider the Boolean expression.

$$(x_1 \vee x_2) \wedge (x_1' \vee x_2') \wedge (x_2 \vee x_3)'$$

Or the Boolean algebra $(\{0, 1\} \wedge, \vee, ')$.

If $x_1 = 0, x_2 = 1, x_3 = 0$, then the above expression becomes

$$(0 \vee 1) \wedge (0' \vee 1') \wedge (1 \vee 0)' = 1 \wedge 1 \wedge 0 = 0$$

Observe that Boolean expressions can be simplified to a great extent. For example, consider the following expression.

$$(a \vee b \vee c \vee d') \wedge (a \vee b \vee c \vee d) \wedge (a \vee b \vee b') \wedge (a \vee b)$$

To simplify this let

$$a \vee b \vee c = x.$$

Then

$$\begin{aligned}(a \vee b \vee c \vee d') \wedge (a \vee b \vee c \vee d) &= (x \vee d') \wedge (x \vee d)\\ &= x \wedge (x \vee d) \vee d' \wedge (x \vee d) \text{ (distributive)}\\ &= x \vee d' \wedge x \vee d' \wedge d \text{ (absorption and distributive)}\\ &= (x \vee d' \wedge x) \vee 0 \text{ (associative and law of bounds)}\\ &= x \text{ (absorption and law of bounds)}\end{aligned}$$

Thus the given expression is given by

$$(a \vee b \vee c) \wedge (a \vee b \vee c') \wedge (a \vee b')$$

Put $a \vee b = y$. Then

$$(a \vee b \vee c) \wedge (a \vee b \vee c') = (y \vee c) \wedge (y \vee c')$$
$$= y \wedge (y \vee c) \vee c \wedge (y \vee c') \text{ (distributive)}$$
$$= y \vee c \wedge y \vee c \wedge c' \text{ (absorption and distributive)}$$
$$= (y \vee c \wedge y) \vee 0 \text{ (associative and law of bounds)}$$
$$= y \text{ (absorption and law of bounds)}$$

Hence the given expression is,

$$(a \vee b) \wedge (a \vee b')$$
$$= a \wedge (a \vee b') \vee b \wedge (a \vee b') \text{ (distributive)}$$
$$= a \vee b \wedge a \vee b \wedge b' \text{ (absorption and distributive)}$$
$$= (a \vee b \wedge a) \vee 0 \text{ (associative and law of bounds)}$$
$$= a \text{ (absorption and law of bounds)}$$

The algebra of logic (the collection of statements) is a Boolean algebra with respect to disjunction, conjunction and negation and in this case the universal bounds are tautology and the contradiction.

We can apply the laws of Boolean algebra to algebra of logic so that we can simplify complicated statements into simple ones. We can even avoid drawing truth tables which are sometimes complicated. For example, take

Let $p = (\neg p \wedge \neg q) \vee (p \vee q) \wedge (\neg p \vee q) \wedge q)$

Since $\neg(p \vee q) = \neg p \wedge \neg q$ (by De Morgan laws)

We have

$$(\neg p \wedge \neg q) \vee (p \vee q) = \neg(p \vee q) \vee (p \vee q)$$
$$= \text{tautology (by laws of complementation)}$$

Hence $p =$ tautology $\wedge ((\neg p \wedge p) \wedge q)$

$$= (\neg p \wedge p) \wedge q \text{ (law of bounds)}$$
$$= q \text{ (absorption law)}$$

The reader is advised to verify this by truth table.

We shall now discuss the application of Boolean algebra to switching circuits.

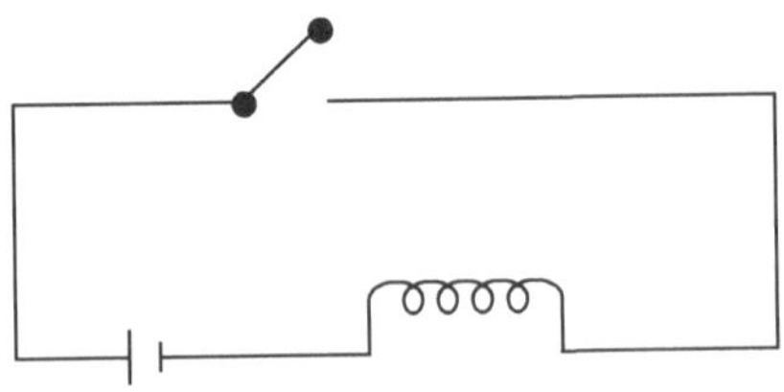

Fig. 1

A switching circuit consists of a battery, a bulb, some switches and connecting wires. It has two states: on or off. When the switch is on, the current flows and the bulb shines and when the switch is off, the current does not flow and the bulb does not shine. The value 1 or the symbol T is associated with the 'on' state of the switch and the value o or the letter F denotes the 'off' state of the switch. The switches in a circuit are denoted by the alphabets $a, b, c, \ldots, x, \ldots$ etc. Note that when two switches in a circuit are both 'on' or both 'off ', then we denote them by the same letter and if one switch is 'on' and the other is 'off', them we denote one by a and the other by a' (Complement of a).

8.5 SERIES AND PARALLEL SWITCHES

Two switches in circuit can be connected either in series or in parallel.

The following two figures show the series and parallel connections.

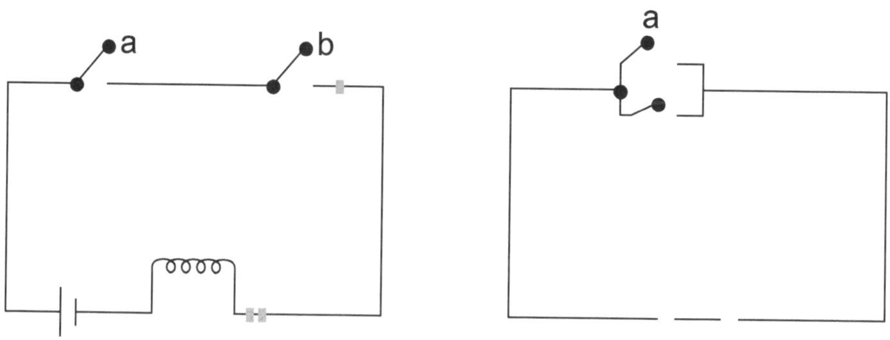

Fig. 2

The series and parallel connections of two switches a and b can be denoted by $a \wedge b$, and $a \vee b$ respectively and this can be represented in the tabular form,

Series Connections

a	a	$a \wedge b$
1	1	1
0	1	0
0	0	0
1	0	0

Parallel Connections

a	b	$a \vee b$
1	1	1
1	0	1
0	0	0
0	1	1

Simplification of switching circuits.

8.5.1 *Example*

Simplify the following circuit.

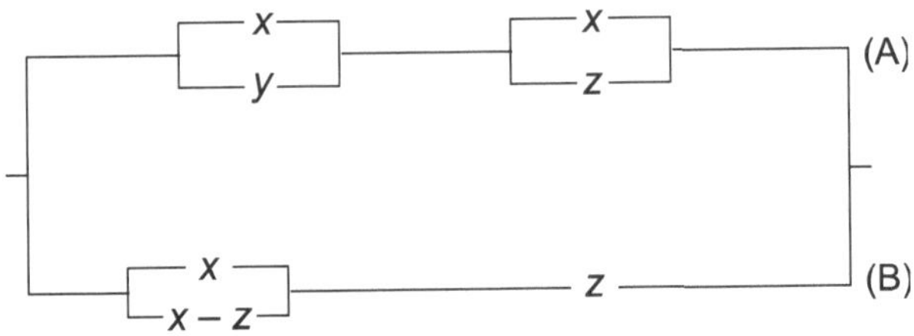

Fig. 3

We will first calculate the Boolean function which represents the above circuit. The circuit has two branches (A) and (B).

For (A):
The switches x and y are parallel, hence it is $x \vee y$. The switches x and z are parallel, hence it is $x \vee z$,

The above two are in series, hence the representation of (A) is,

$$(x \vee y) \wedge (x \vee z).$$

For (B):
x and z are in series, hence we have $x \wedge z$, the above is parallel to x, hence $x \vee (x \wedge z), x \vee (x \wedge z)$ is in series with x.

So we have

$$\big(x \vee (x \wedge z)\big) \vee z.$$

(A) and (B) are parallel. Hence the Boolean function for the whole circuit is,

$$(x \vee y) \wedge (x \vee z) \vee \big(x \vee (x \wedge z)\big) \wedge z$$
$$= x \wedge (x \vee z) \vee y \wedge \big(x \vee z) \vee (x \wedge z)\big)$$
$$= x \vee (x \wedge z) \wedge y \vee \big(x \vee z) \wedge (x \vee z)\big)$$
$$= x \vee (x \wedge z) \vee (y \vee z)$$
$$= x \vee (y \wedge z)$$

8.5.2 *Example*

Simplify the following network.

There are two parts in this circuit, the first part has representation,

$$x \vee y \ \vee (a \wedge b \wedge c)$$

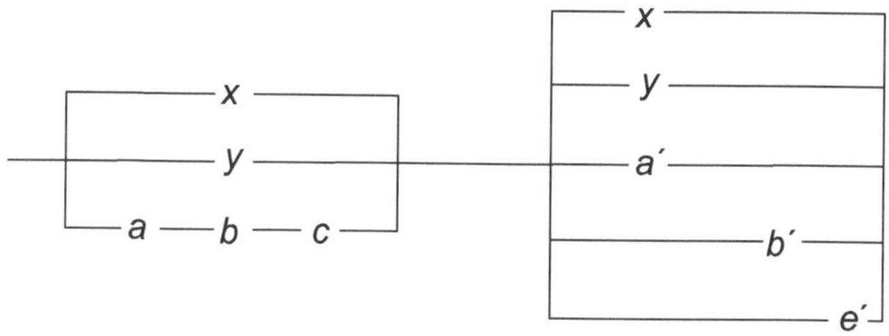

Fig. 4

and the second has the representation,

$$(x \vee y \vee (a \wedge b \wedge c)) \wedge (x \vee y \vee a' \vee b' \vee c')$$
$$= (x \vee y) \vee (a \wedge b \wedge c)) \wedge (x \vee y \vee (a \vee b \vee c)')$$
$$= (x \vee y) \wedge (x \vee y) \vee (a \wedge b \wedge c) \wedge (x \vee y) \wedge (x \vee y) \wedge (a \wedge b \wedge c)' \vee (a \wedge b \wedge c) \wedge (a \wedge b \wedge c)$$
$$= [(x \vee y) \vee (a \wedge b \wedge c) \vee (x \vee y)] \vee (x \vee y) \wedge (a \wedge b \wedge c)' \vee 0$$
$$= (x \vee y) \vee (x \vee y) \wedge (a \wedge b \wedge c)'$$
$$= (x \vee y)$$

The circuit represented by $x \vee y$ is

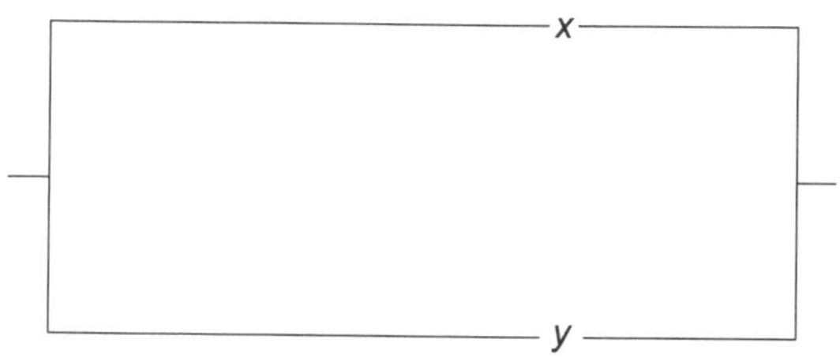

and this is the circuit to which the given network is equivalent.

EXERCISES

1. Define a Boolean algebra. Show that, in a Boolean algebra, every element has a unique complement. Also show that if x and y arc elements of a Boolean algebra then the identities:

   $(x \wedge x)' = x' \vee y'$ and $(x \vee y)' = x' \wedge y'$ are true.

2. Let L be a lattice, if x, y, z are elements of L, verify the following:

   $x \wedge x = x, x \vee x = x, x \wedge y = y \wedge x, x \vee y = y \vee x$

   $x \wedge (y \wedge z) = (x \wedge y) \wedge z, x \vee (y \vee z) = (x \vee y) \vee z,$

   $(x \wedge y) \vee x = x, (x \vee y) \wedge x = x.$

3. Let L be a nonempty set in which two operations $\wedge$ and $\vee$ are defined which satisfy the following conditions:

$$x \wedge x = x, \qquad x \vee x = x;$$
$$x \wedge y = y \wedge x, \qquad x \vee y = y \vee x;$$
$$x \wedge (y \wedge z) = (x \wedge y) \wedge z, \qquad x \vee (y \vee z) = (x \vee y) \vee z;$$
$$(x \wedge y) \wedge x = x, \qquad (x \vee y) \wedge x = x.$$

Show that L is a partially ordered set in which any two elements have a greatest lower bound and a least upper bound.

CHAPTER 9

Theory of Fuzzy Sets

Advances in science and technology have made our modern society very complex and hence decision process has become increasingly vague and hard to analyze. Human brain has some special characteristic by which it is possible to learn and reason in a vague and fuzzy environment. It has the ability to arrive at decision based on imprecise, qualitative data in contrast to formal mathematics and formal logic which demands precise and quantitative data. Modern computers have over taken the brain's capacity to some extent, but still the human brain holds its charm by its ability to take decisions based on imprecise and qualitative data. Undoubtedly, in many areas of cognition human intelligence excels, the computer intelligence of today, and the development of tools capable of handling humanistic type of problem.

We have sufficient mathematical tools and computer based technology for analyzing and solving the problems embodied in deterministic and uncertain environment. Here uncertainty may arise from the probabilistic behavior of certain physical phenomena in mechanistic systems or due to qualitative data. We know the important role that vagueness and in exactitude play in human decision making, but until 1965 we did not know how the vagueness arising from subjectivity (which is inherent in human thought processes) can be modelled and analyzed.

John Von Neumann gave a serious thought in this direction during early fifties. John Von Neumann was born in Budapest, Hungary in 1903. From his childhood he had shown a remarkable ability and interest in scientific matters. In 1943 he became convinced and tried to convince others in many varied fields, that numerical calculations done on fast electronic computing devices, would substantially facilitate the solutions of many difficult, unsolved scientific problems. After the controversy with a group of selected engineers and mathematicians, Neumann built, at the Institute for advanced study, an experimental electronic calculator, popularly known as the JONIAC which eventually became the pilot model for similar machines all over the country. They tried to design the machine by initiating some of the known operations of the brain. For this he started reading neurology.

Early in 1955 he was invited by Yale University to give the Silliman lectures during the spring term of 1956. In Silliman lectures he wanted to give the improved idea of copying an extremely simplified model of the human brain for man-made machines. But in August 1955 he had developed severe pain in his left shoulder and it was diagnosed to be bone cancer. In spite of many efforts, he could not finish writing his planned lectures and died on February 8, 1957. But in order to pay respect to the departed soul, Yale University published his incomplete manuscript with a preface written by his wife. The last section of his unfinished manuscript was "The Language of Mathematics is not the Language of Human Beings".

Later in early sixties Professor L.A. Zadeh of the University of California, Berkeley thought in this direction. He asked "Can computer be made to think like human beings?"

In 1965, Prof. Zadeh laid the foundation of fuzzy set theory. Fuzzy set theory is the concept and technique which lay a form of mathematical precision to human thought processes that in many ways were imprecise and ambiguous by the standards of classical mathematics. The fuzzy theory has blossomed into a many faceted field of inquiry, drawing on and contributing to a wide spectrum of areas ranging from para mathematics to human perception and judgement. Fuzzy sets have been suggested for handling the imprecise real word problems by using truth values ranging between the usual "true" and "false".

There are classes of objects encountered in the real physical world which do not have precisely defined criteria of membership. For example, consider the collection of all small natural numbers or the class of all students with brilliant academic career or the collection of all tall men etc. All the classes defined above do not constitute sets in the usual mathematical sense. Yet, the fact remains that such imprecisely defined "classes" play an important role in human thinking particularly in the domains of pattern recognition, communication of information and abstraction.

L.A. Zadeh in 1965 explored a preliminary concept for these classes of ambiguities. He introduced the fuzzy set which is defined as a class with continnum of grades of membership. The behavior of fuzzy sets is similar in many senses with that of the ordinary sets but is more general than the later. It has much wider scope of applicability, particularly in the fields of pattern classification and information processing. It provides a natural way of dealing with problems in which the source of imprecision is the absence of sharply defined criteria of class of membership rather than the presence of random variables. Thus, a set can be called fuzzy set (but not vice-versa). Starting with the concepts in the set theory we will see how they lead to a generalized theory of fuzzy sets.

9.1 THE CONCEPT OF MEMBERSHIP

Let E be a set and $A \subseteq E$. Recall that 'an element x of E is a member of A' is indicated by using the symbol $\in$; that is $x \in A$ In order to indicate this membership one can use another concept also: the characteristic function μ_A from E taking values in $\{0.1\}$ is defined by,

$$\mu_A(x) = 1 \text{ if } x \in A$$
$$= 0 \text{ if } x \notin A$$

Let A and be B any two subsets of E.

We have

$$\mu_A(x) = 1 \text{ if } x \in A$$
$$= 0 \text{ if } x \notin A$$
$$\mu_B(x) = 1 \text{ if } x \in B$$
$$= 0 \text{ if } x \notin B$$
$$\mu_{A \cap B}(x) = 1 \text{ if } x \in A \cap B$$
$$= 0 \text{ if } x \notin A \cap B$$
$$\mu_{A \cup B}(x) = 1 \text{ if } x \in A \cup B$$
$$= 0 \text{ if } x \notin A \cup B$$

This allows us to write

$$\mu_{A \cap B} = \mu_A(x) \,.\, \mu_B(x)$$
$$\mu_{A \cup B} = \mu_A(x) + \mu_B(x)$$

where the operations $\cdot$ and $+$ correspond to tables in Figures 1 and 2 respectively and are called the Boolean product and Boolean sum.

·	0	1
0	0	0
1	0	1

Figure 1

+	0	1
0	0	1
1	1	1

Figure 2

Let A' be the complement of A with respect to E. Then,

$$\mu_A(x) = 1 \text{ if } x \in A$$
$$= 0 \text{ if } x \notin A$$
$$\mu_{A'}(x) = 1 \text{ if } x \in A'$$
$$= 0 \text{ if } x \notin A'$$

Further $A \cap A' = \emptyset, A \cup A' = E$. Thus we can write

$$\mu_{A'}(x) = 1 - \mu_A(x)$$

9.1.1 *Example*

1. Let $E = \{x_1, x_2, x_3, x_4, x_5\}$ and $A = \{x_1, x_2\}$
   We can represent the set A by accompanying the elements of E with their characteristic function values:
   $A = \{(x_1, 1), (x_2, 1), (x_3, 0), (x_4, 0), (x_5, 0)\}$
   Similarly, A' can be represented by,
   $A' = \{(x_1, 0), (x_2, 0), (x_3, 1), (x_4, 1), (x_5, 1)\}$
2. Let
   $B = \{(x_1, 1), (x_2, 0), (x_3, 1), (x_4, 0), (x_5, 1)\}$
   Then
   $A \cap B = \{(x_1, 1, 1), (x_2, 1, 0), (x_3, 0, 0), (x_4, 0, 0), (x_5, 0, 0)\}$
   $= \{(x_1, 0), (x_2, 0), (x_3, 0), (x_4, 0), (x_5, 0)\}$
   $A \cup B = \{(x_1, 1 + 0), (x_2, 0 + 0), (x_3, 0 + 0), (x_4, 0 + 0), (x_5, 0 + 0)\}$
   $= \{(x_1, 1), (x_2, 1), (x_3, 0), (x_4, 0), (x_5, 0)\}$

9.2 THE CONCEPT OF FUZZY SUB SET

Let E be a set. Let M be a totally ordered set (in particular M could be the interval [0, 1]). Then a fuzzy subset A of E is a set of ordered pairs,

$$\{(x, \mu_A(x))\}, \forall x \in E$$

where μ_A is a function defined on E which takes its values in M. M is called a membership set and μ_A is called a membership or characteristic function. $\mu_A(x)$ is called the grade or degree or level of member-ship of x in A. If $M = \{0, 1\}$ then 'fuzzy subset' is an 'ordinary subset'.

The notion of fuzzy subsets is a generalization of the ordinary subsets to the case where the membership of elements are not clear cut and allows one to study, using mathematical structures, imprecise concept.

A set or a subset is denoted by $A, B, C, X, Y, Z, \ldots\ldots$ A fuzzy subset will be denoted by $\underset{\sim}{A}, \underset{\sim}{B}, \underset{\sim}{C}, \underset{\sim}{X}, \underset{\sim}{Y}, \underset{\sim}{Z}$ …… If all the subsets under consideration are ordinary, then one may, suppress the small supplementary symbol ~. Membership and non-membership are denoted by $\in$ and $\notin$. Fuzzy membership and fuzzy non-membership will be denoted by $\underset{\sim}{\in}$ and $\underset{\sim}{\notin}$.

In certain cases where the membership set M is the closed interval $[0, 1]$ it may be convenient to represent the membership by accompanying the symbol $\in$ by a number from $[0, 1]$ placed beneath it. Thus

$x \underset{1}{\in} \underset{\sim}{A}$ means $x \in \underset{\sim}{A}$

$x \underset{0}{\in} \underset{\sim}{A}$ means $x \notin \underset{\sim}{A}$

$x \underset{0,8}{\in} \underset{\sim}{A}$ means $x \underset{\sim}{\in} \underset{\sim}{A}$ with degree 0, 8

$x \underset{0,2}{\in} \underset{\sim}{A}$ means $x \underset{\sim}{\in} \underset{\sim}{A}$ with degree 0.2.

and so on.

Let $E = \{x_1, x_2, x_3, x_4, x_5\}$

and let $M = [0, 1]$.

Then $A = \{ (x_1 / 0.2), (x_2 / 0), (x_3 / 0.3), (x_4 / 1), (x_5 / 0.8)\}$

where x_i is an element of the reference set E and where the number placed after the bar is the value of the characteristic function for the element, is a fuzzy subset of E and is denoted by,

$$\underset{\sim}{A} \subset E$$

3. Let $E = \{a, b, c, d, e, f\}$ and $M = \{0, \frac{1}{2}, 1\}$.

   Then $A = \{(a/0), (b/1), (c/1/2), (d/0), (e/1/2), (f/0)\}$ is a fuzzy subset of E.

4. Let N be the set of all natural numbers:

   $N - \{1, 2, 3, 4, \ldots\ldots.\}$ and $M = [0, 1]$. Then

   $A = \{(0|1), (1|0,8), (2|0.6), (3|0.4), (4|0.2), (5|0), (6|0), \ldots\ldots\ldots\ldots\}$

   is a fuzzy subset of "small" natural numbers.

5. Let $E = \{0, 1, 2, 3, 4, 5, 6, 7, 8, 9\}$ and $M = [0, 1]$. Then

$$A = \{(0|0), (1|0.2), (2|0.3), (3|0), (4|1), (5|1), (6|0.8), (7|0.9), (8|0), (9|0)\}$$

   is a fuzzy subset of E.

9.3 SIMPLE OPERATIONS ON FUZZY SUB SETS

Inclusion: Let E be a set and M its associated membership set. Let $\underset{\sim}{A}$ and $\underset{\sim}{B}$ be two fuzzy subsets of E. A is said to be included in B written as $\underset{\sim}{A} \subset \underset{\sim}{B}$ or $\underset{\sim}{A} \underset{\sim}{\subseteq} \underset{\sim}{B}$ if,

$$\mu_{\underset{\sim}{A}}(x) \le \mu_{\underset{\sim}{B}}(x) \text{ for all } x \in E.$$

A is said to be strictly included in B written $\underset{\sim}{A} \subset\subset \underset{\sim}{B}$ or $\underset{\sim}{A} \underset{\sim}{\subseteq}\subset \underset{\sim}{B}$ if the above inequality is strict for at least one $x \in E$.

6. Let $E = \{x_1, x_2, x_3, x_4\}, M = [0, 1]$.

   $\underset{\sim}{A} = \{(x_1| \, 0.4)(x_2| \, 0.2), (x_3| \, 0), (x_4 \, |1)\}$

   $\underset{\sim}{B} = \{(x_1| \, 0.3)(x_2| \, 0), (x_3| \, 0), (x_4 \, |0)\}$

   One has $\underset{\sim}{B} \subset \underset{\sim}{A}$ since $0.3 < 0.4, 0 < 0.2, 0 = 0, 0 < 1$.

   One can write

   $$E = \{(x_1| \, 1)(x_2| \, 1), (x_3| \, 1), (x_4 \, |1)\}$$

   Thus, $E \underset{\sim}{\subseteq} E$ i.e., E is included in itself in the sense of the theory of fuzzy subsets.

Equality: Let E be a set, M its associated membership set, and $\underset{\sim}{A}$ and $\underset{\sim}{B}$ two fuzzy subsets of E. $\underset{\sim}{A}$ and $\underset{\sim}{B}$ are said to be equal if and only if,

$$\mu_{\underset{\sim}{A}}(x) = \mu_{\underset{\sim}{B}}(x) \text{ for all } x \in E.$$

This will be written as $\underset{\sim}{A} = \underset{\sim}{B}$. If for at least one $x \in E$, $\mu_{\underset{\sim}{A}}(x) \ne \mu_{\underset{\sim}{B}}(x)$ then $\underset{\sim}{A}$ and $\underset{\sim}{B}$ are said to be unequal and written as, $\underset{\sim}{A} \ne \underset{\sim}{B}$.

Complementation: Let E be a set and $M = [0, 1]$ its associated membership set. Let $\underset{\sim}{A}$ and $\underset{\sim}{B}$ be two fuzzy subsets of E. $\underset{\sim}{A}$ is the complement of $\underset{\sim}{B}$ or $\underset{\sim}{B}$ is the complement of $\underset{\sim}{A}$.

$$\mu_{\underset{\sim}{B}}(x) = 1 - \mu_{\underset{\sim}{A}}(x) \text{ for all } x \in E.$$

This will be denoted by $\underset{\sim}{B} = \underset{\sim}{A}'$ or $\underset{\sim}{A} = \underset{\sim}{B}'$.

One obviously has $(\underset{\sim}{A}')' = \underset{\sim}{A}$

7. Let $E = \{x_1, x_2, x_3, x_4, x_5, x_6\}, M = [0, 1]$.

   $\underset{\sim}{A} = \{(x_1| \, 0.13), (x_2| \, 0.61), (x_3| \, 0), (x_4 \, |0), (x_5 \, |1), (x_6 \, |0.03)\}$

   $\underset{\sim}{B} = \{(x_1| \, 0.87), (x_2| \, 0.39), (x_3| \, 1), (x_4 \, |1), (x_5 \, |0), (x_6 \, |0.97)\}$

   Then $\underset{\sim}{A}' = \underset{\sim}{B}$

Intersection: Let E be a set and $M = [0, 1]$ its associated membership set and let $\underset{\sim}{A}$ and $\underset{\sim}{B}$ be two fuzzy subsets of E. Then the intersection of $\underset{\sim}{A}$ and $\underset{\sim}{B}$ denoted by $\underset{\sim}{A} \cap \underset{\sim}{B}$ is defined as the largest fuzzy subset contained at the same time in $\underset{\sim}{A}$ and $\underset{\sim}{B}$. That is

$$\mu_{\underset{\sim}{A} \cap \underset{\sim}{B}}(x) = \min\{\mu_{\underset{\sim}{A}}(x),\ \mu_{\underset{\sim}{B}}(x) \text{ for all } x \in E.$$

8. Let $\underset{\sim}{A} = \{(x_1 |\, 0.2), (x_2 |\, 0.7), (x_3 |\, 1), (x_4 \,|0), (x_5 |\, 0.5)\}$

   $\underset{\sim}{B} = \{(x_1 |\, 0.5), (x_2 |\, 0.3), (x_3 |\, 1), (x_4 \,|0.1), (x_5 |\, 0.5)\}$

   $\underset{\sim}{A} \cap \underset{\sim}{B} = \{(x_1 |\, 0.2), (x_2 |\, 0.3), (x_3 |\, 1), (x_4 \,|0), (x_5 |\, 0.5)\}$

Union: Let E be a set and $M = [0, 1]$ its associated membership set and let $\underset{\sim}{A}$ and $\underset{\sim}{B}$ be two fuzzy subsets of E. The union of $\underset{\sim}{A}$ and $\underset{\sim}{B}$ denoted by $\underset{\sim}{A} \cup \underset{\sim}{B}$ is defined as the smallest fuzzy subset that contains both $\underset{\sim}{A}$ and $\underset{\sim}{B}$. That is,

$$\mu_{\underset{\sim}{A} \cup \underset{\sim}{B}}(x) = \max\{\mu_{\underset{\sim}{A}}(x),\ \mu_{\underset{\sim}{B}}(x)\} \text{ for all } x \in E.$$

In Example 8 we have

Let $\underset{\sim}{A} \cup \underset{\sim}{B} = \{(x_1 |\, 0.5), (x_2 |\, 0.7), (x_3 |\, 1), (x_4 \,|0.1), (x_5 |\, 0.5)\}$

Difference: Let E be a set and $M = [0, 1]$ its associated membership set and let $\underset{\sim}{A}$ and $\underset{\sim}{B}$ be two fuzzy subsets of E. The difference of $\underset{\sim}{A}$ and $\underset{\sim}{B}$ is defined by the relation,

$$\underset{\sim}{A} - \underset{\sim}{B} = \underset{\sim}{A} \cap \underset{\sim}{B}'$$

Considering Example 8 we have,

8. $\underset{\sim}{A} - \underset{\sim}{B}' = \{(x_1 |\, 0.2), (x_2 |\, 0.7), (x_3 |\, 0), (x_4 \,|0), (x_5 |\, 0.5)\}$

   Note that, except in particular cases,

   $\underset{\sim}{A} - \underset{\sim}{B} \neq \underset{\sim}{B} - \underset{\sim}{A}$

9.4 VISUAL REPRESENTATION OF FUZZY SETS

We know that ordinary subsets can be represented by Venn diagram. For fuzzy subsets one can also construct a visual representation.

Consider a rectangle (Figure 1) with the values $\mu_{\underset{\sim}{A}}(x)$ as ordinate and as abscissa the elements of E in an arbitrary order (if there is some total order in E that order has to be taken). In Figure 1 the membership of each element is represented by its ordinate. The shaded part represents the fuzzy subsets

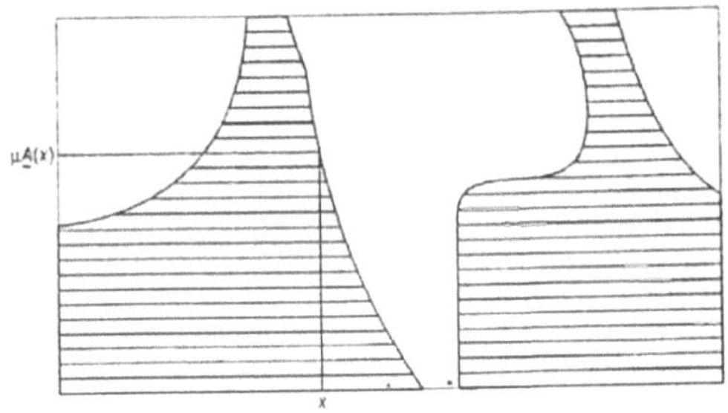

Figure 1

$\underset{\sim}{A} \subset E$. With this representation one can visualize the various simple operations such as: Complementation, union, intersection etc. for fuzzy subsets.

9.5 FUZZY SUBSET FOR E AND M FINITE

Let E and M be finite. We first recall the definition of the set of all subsets (or power set) of a set by considering a simple example. Let $E = \{x_1, x_2, x_3\}$.

Then the power set of $E = \varphi(E)$, is represented as follows:

$$\varphi(E) = \{\emptyset, \{x_1\}, \{x_2\}, \{x_3\}, \{x_1, x_2\}, \{x_1, x_3\} \{(x_2, x_3), E\}$$

There are $2^3 = 8$ elements in this set. More generally if,

$$E = \{x_1, x_2, x_3, \dots \dots \dots .., x_n\}$$

Has n elements, $\varphi(E)$ will have 2^nelements.

For a fuzzy subset, the power set or "the set of fuzzy subsets" is presented in slightly different manner. Let

Let $E = \{x_1, x_2\}$ and $M = [0, 1/2, 1]$.

Then the set of fuzzy subsets, $\varphi(\underset{\sim}{E})$ will be

$$\varphi(\underset{\sim}{E}) = \{(x_1| \ 0), (x_2| \ 0)\}, \{(x_1| \ 0), (x_2 \ |0.5)\}$$
$$\{(x_1| \ 0.5), (x_2| \ 0)\}, \{(x_1| \ 0.5), (x_2 \ |0.5)\}$$
$$\{(x_1| \ 0), (x_2| \ 1)\}, \{(x_1| \ 1), (x_2 \ |0)\}$$
$$\{(x_1| \ 1), (x_2| \ 0.5)\}, \{(x_1| \ 0.5), (x_2 \ |1)\}$$
$$\{(x_1(| \ 1), (x_2| \ 1)\}$$

More generally, if

$$Card\ E = n \text{ and } Card\ M = m$$

where card means "cardinality of", that is, the number of elements for a finite set, then

$$\text{card}\ (\varphi(\underset{\sim}{E})) = m^n$$

The set $\varphi(\underset{\sim}{E})$ contains 2^n ordinary subsets. Note that one always has $\varphi(E) \subset \varphi(\underset{\sim}{E})$.

9.6 PROPERTIES OF FUZZY SUBSET

Recall that the power set E of an ordinary set has the following properties: for $A \subset E, B \subset E, C \subset E$ and we have,

1. $A \cap B = B \cap A,$
2. $A \cup B = B \cup A,$
3. $(A \cap B) \cap C = A \cap (B \cap C),$
4. $(A \cup B) \cup C = A \cup (B \cup C),$
5. $A \cap A = A,$
6. $A \cup A = A,$
7. $A \cap (B \cup C) = (A \cap B) \cup (A \cap C),$
8. $A \cup (B \cap C) = (A \cup B) \cap (A \cup C),$
9. $A \cap A' = \emptyset,$
10. $A \cup A' = E,$
11. $A \cap \varphi = \varphi$
12. $A \cup \emptyset = A,$
13. $A \cap E = A,$
14. $A \cup E = E,$
15. $(A')' = A,$
16. $(A \cap B)' = A' \cup B',$
17. $(A \cup B)' = A' \cap B'$

    If $\underset{\sim}{A}, \underset{\sim}{B}, \underset{\sim}{C}$ are fuzzy subsets of E all the properties 1–17 hold except 9 and 10, these two hold for ordinary subsets only. For fuzzy subsets we have.

18. $\underset{\sim}{A} \cap \underset{\sim}{B} = \underset{\sim}{B} \cap \underset{\sim}{A}$
19. $\underset{\sim}{A} \cup \underset{\sim}{B} = \underset{\sim}{B} \cup \underset{\sim}{A}$
20. $(\underset{\sim}{A} \cap \underset{\sim}{B}) \cap \underset{\sim}{C} = \underset{\sim}{A} \cap (\underset{\sim}{B} \cap \underset{\sim}{C})$

21. $(\underset{\sim}{A} \cup \underset{\sim}{B}) \cup \underset{\sim}{C} = \underset{\sim}{A} \cup (\underset{\sim}{B} \cup \underset{\sim}{C})$
22. $\underset{\sim}{A} \cap \underset{\sim}{A} = \underset{\sim}{A}$
23. $\underset{\sim}{A} \cup \underset{\sim}{A} = \underset{\sim}{A}$
24. $\underset{\sim}{A} \cap (\underset{\sim}{B} \cup \underset{\sim}{C}) = (\underset{\sim}{A} \cap \underset{\sim}{B}) \cup (\underset{\sim}{A} \cap \underset{\sim}{C})$
25. $\underset{\sim}{A} \cup (\underset{\sim}{B} \cap \underset{\sim}{C}) = (\underset{\sim}{A} \cup \underset{\sim}{B}) \cap (\underset{\sim}{A} \cup \underset{\sim}{C})$
26. $\underset{\sim}{A} \cap \emptyset = \emptyset$ where $\emptyset$ is the ordinary set such that $\mu_\phi(x_i) = 0$ for all $x_i \in E$.
27. $\underset{\sim}{A} \cup \emptyset = \underset{\sim}{A}$
28. $\underset{\sim}{A} \cap \text{E} \underset{\sim}{A}$, where E is such that $\mu_E(x_i) = 1$ for all $x_i \in E$. (i.e. the (ordinary) universal reference set).
29. $\underset{\sim}{A} \cup E = E$
30. $(\underset{\sim}{A}')' = \underset{\sim}{A}$
31. $(\underset{\sim}{A} \cap \underset{\sim}{B})' = \underset{\sim}{A}' \cup \underset{\sim}{B}'$
32. $(\underset{\sim}{A} \cup \underset{\sim}{B})' = \underset{\sim}{A}' \cap \underset{\sim}{B}'$

9.7 ALGEBRAIC PRODUCT AND SUM OF TWO FUZZY SUBSETS

Let E be a set and $M = [0, 1]$ its associated membership set. Let $\underset{\sim}{A}$ and $\underset{\sim}{B}$ be two fuzzy subsets of E. We define the algebraic product of $\underset{\sim}{A}$ and $\underset{\sim}{B}$, denoted by $\underset{\sim}{A} \cdot \underset{\sim}{B}$ and algebraic sum of $\underset{\sim}{A}$ and $\underset{\sim}{B}$ denoted by $\underset{\sim}{A} \hat{+} \underset{\sim}{B}$ in the following ways:

$$\bar{\mu}_{\underset{\sim}{A} \cdot \underset{\sim}{B}}(x) = \mu_{\underset{\sim}{A}}(x) \cdot \mu_{\underset{\sim}{B}}(x), \ \forall x \in E$$

$$\mu_{\underset{\sim}{A} \hat{+} \underset{\sim}{B}}(x) = \mu_{\underset{\sim}{A}}(x) + \mu_{\underset{\sim}{B}}(x) - \mu_{\underset{\sim}{A}}(x) \cdot \mu_{\underset{\sim}{B}}(x), \ \forall x \in E$$

9. Let

$$\underset{\sim}{A} = \{(x_1|\, 0.2), (x_2|\, 0.7), (x_3|\, 1), (x_4\, |0), (x_5|\, 0.5)\}$$

$$\underset{\sim}{B} = \{(x_1|\, 0.5), (x_2|\, 0.3), (x_3|\, 1), (x_4\, |0.1), (x_5|\, 0.5)\}$$

Then

$$\underset{\sim}{A} \cdot \underset{\sim}{B} = \{(x_1|\, 0.10), (x_2|\, 0.21), (x_3|\, 1), (x_4\, |0), (x_5|\, 0.25)\}$$

$$\underset{\sim}{A} \hat{+} \underset{\sim}{B} = \{(x_1|\, 0.60), (x_2|\, 0.79), (x_3|\, 1), (x_4\, |0.1), (x_5|\, 0.75)\}$$

Note that if $M = \{0, 1\}$ i.e. if we are discussing about ordinary subsets, then

$$A \cap B = A \cdot B, A \cup B = A \overset{\wedge}{+} B$$

One can verify the following properties on the power set of fuzzy subsets with the two operations $\cdot$ $\overset{\wedge}{+}$ and. It may be noted here properties similar to 5 and 6, 7 and 8 (distributivity), 9 and 10 with $\cup$ and $\cap$ replaced by $\overset{\wedge}{+}$ and, ordinary subsets replaced by fuzzy subsets are not true. What are true are only the following:

33. $\underset{\sim}{A} \cdot \underset{\sim}{B} = \underset{\sim}{B} \cdot \underset{\sim}{A}$

34. $\underset{\sim}{A} \overset{\wedge}{+} \underset{\sim}{B} = \underset{\sim}{B} \overset{\wedge}{+} \underset{\sim}{A}$

35. $(\underset{\sim}{A} \cdot \underset{\sim}{B}) \cdot \underset{\sim}{C} = \underset{\sim}{A} \cdot (\underset{\sim}{B} \cdot \underset{\sim}{C})$

36. $(\underset{\sim}{A} \overset{\wedge}{+} \underset{\sim}{B}) \overset{\wedge}{+} \underset{\sim}{C} = \underset{\sim}{A} \overset{\wedge}{+} (\underset{\sim}{B} \overset{\wedge}{+} \underset{\sim}{C})$

37. $\underset{\sim}{A} \cdot \emptyset = \emptyset$

38. $\underset{\sim}{A} \overset{\wedge}{+} \emptyset = \underset{\sim}{A}$

39. $\underset{\sim}{A} \cdot E = \underset{\sim}{A}$

40. $\underset{\sim}{A} \overset{\wedge}{+} E = E$

41. $(\underset{\sim}{A} \cdot \underset{\sim}{B})' = (\underset{\sim}{A})' \overset{\wedge}{+} (\underset{\sim}{B})'$

42. $(\underset{\sim}{A} \overset{\wedge}{+} \underset{\sim}{B})' = (\underset{\sim}{A})' \cdot (\underset{\sim}{B})'$

EXERCISES

1. Find $\underset{\sim}{A} \cup \underset{\sim}{B}$, $\underset{\sim}{A} \cap \underset{\sim}{B}$, $\underset{\sim}{A} - \underset{\sim}{B}$, $\underset{\sim}{B} - \underset{\sim}{A}$ in the following cases:

   (i) $\underset{\sim}{A} = \{(x_1 | 0.2), (x_2 | 0.8), (x_3 | 0.6)\}$

   $\underset{\sim}{B} = \{(x_1 | 0.4), (x_2 | 0.6), (x_3 | 1.0)\}$.

   (ii) $\underset{\sim}{A} = \{(x_1 | 1.0), (x_2 | 0.0), (x_3 | 0.5), (x_4 | 0.7)\}$

   $\underset{\sim}{B} = \{(x_1 | 0.8), (x_2 | 1.0), (x_3 | 0.0), (x_4 | 0.6)\}$

2. If $E = \{x_1, x_2, x_3, x_4\}$ and $M = [0, 1]$. Find $\underset{\sim}{A}'$, $\underset{\sim}{B}'$ in 1(i) and (ii).

3. Prove De Morgan's Laws for fuzzy subsets.

4. Find $\underset{\sim}{A} \cdot \underset{\sim}{B}$ and $\underset{\sim}{A} \overset{\wedge}{+} \underset{\sim}{B}$ in 1(i) and (ii).

5. Find the fuzzy power set of the following:
$S = \{x_1, x_2\}, M = \{0, 0.6, 1\}$,
$S = \{x_1, x_2, x_3\}, M = \{0.5, 0.6\}$

Bibliography

1. Das, M.K., *Number Theory*, Orissa State Bureau of Text Book Preparation and Production, Bhubaneswar (1997).
2. Fraleigh, J.B., *A First Course in Abstract Algebra*, Reading, Addison-Wesley 1967 (4th print 1972).
3. Joseph A. Gallian, Contemporary Abstract Algebra, Narosa Publishing House, New Delhi,1999.
4. Halmos, P.R., *Finite Dimensional Vector Spaces*, Princeton University Press (1995).
5. Herstein, I.N., *Topics in Algebra*, Vikas Publishing House Pvt Ltd., New Delhi (1985).
6. Jacobson, N., *Lecturers in Abstract Algebra.* Vols I, II and III, D. Van Nostand Company, Inc. (1965).
7. Kaufman, A., Introduction to the Theory of Fuzzy Subsets, Vol. I, Academic Press, New York.
8. Krishnamurty, V., Mainra, V.P. and Arora, J.L., *An Introduction to Linear Algebra*, Affiliated East-West Press Pvt. Ltd. (1976).
9. Lang, S., *Algebra*, Addition-Wesley Publishing Co. (1972).
10. Nanda, S., *Number Theory*, Allied Publishers Pvt. Ltd., New Delhi (2014).
11. Nanda, M.M., *Modern Algebra*, Orissa State Bureau of Text Book Preparation and Production, Bhubaneswar (1980).
12. Shanti, Narayan and Sat, Pal., *A Text Book of Modern Algebra*, S. Chand and Co., New Delhi (1985).
13. Simmons, G.F., *Introduction to Topology and Modern Analysis*, McGraw-Hill, New York (1963).
14. Tripathy, N., *Theory of Equations*, Orissa State Bureau of Text Book Preparation and Production, Bhubaneswar (1974).
15. Zadeh, L.A., *Fuzzy Sets*, Information and Control, 8 (1965), 338–353.

Index